THE LIVES OF

OCTOPUSES

& THEIR RELATIVES

THE LIVES OF OCTOPUSES

& THEIR RELATIVES

A NATURAL HISTORY OF CEPHALOPODS

Danna Staaf

PRINCETON UNIVERSITY PRESS
PRINCETON AND OXFORD

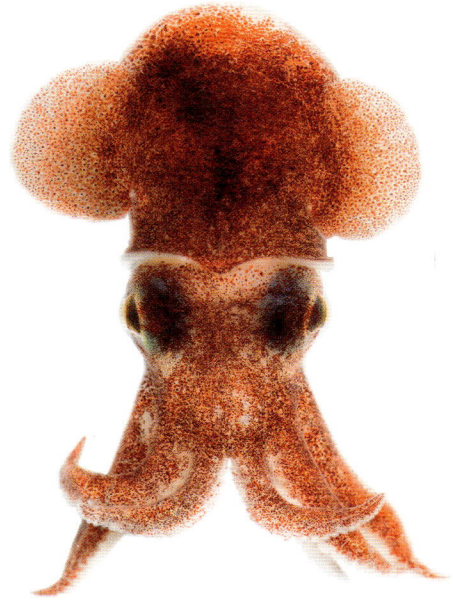

Published by Princeton University Press
41 William Street, Princeton, New Jersey 08540
99 Banbury Road, Oxford OX2 6JX
press.princeton.edu

Library of Congress Control Number 2022947657
ISBN 978-0-691-24430-3
Ebook ISBN 978-0-691-25233-9

Typeset in Bembo and Futura

Printed and bound in Latvia
10 9 8 7 6 5 4 3 2 1

British Library Cataloging-in-Publication Data is available

This book was conceived, designed, and produced by
UniPress Books Limited
Publisher: Nigel Browning
Commissioning editor: Kate Shanahan
Project manager: Kathleen Steeden
Designer & art directon: Wayne Blades
Picture researcher: Julia Ruxton
Illustrator: John Woodcock
Maps: Les Hunt

Cover images: (Front cover) Media Drum World / Alamy Stock
Photo; (back cover and spine) Getty Images

Cover designer: Wanda España

To the wonderfully peculiar and peculiarly wonderful cephalopods
themselves, I am and will remain ever grateful. Since meeting an octopus
in an aquarium at age ten, I've been enamored, delighted, and awed by
these creatures and all their relatives. I'd like to thank Roger Hanlon and
John B. Messenger, whose iconic book on *Cephalopod Behavior* was a
beloved companion as I ventured into the world of research. John
Messenger recently passed away, and the cephalopod world mourns his
loss. Another brilliant book—*Octopus, Squid, and Cuttlefish: A Visual,
Scientific Guide to the Oceans' Most Advanced Invertebrates*, by Roger Hanlon,
Mike Vecchione, and Louise Allcock—taught me so much as I was
working on this project. Knowing that I could never imitate such
a masterpiece, and not wanting to try, this book represents my attempt to
peer at cephalopods through quite a different lens, which I hope I can offer
as a quirky and complementariy work. For the family that has unfailingly
supported me in both science and writing, my love and gratitude are
boundless. Special appreciation to my children, who provided input on
their favorite images for this project. I can't thank Kathy Steeden enough
for her heroic editorial work, Julia Ruxton for tracking down so many
incredible photos, John Woodcock for creating fantastic diagrams, and the
rest of the UniPress team for turning my scribbles into a work of art.

Danna Staaf

CONTENTS

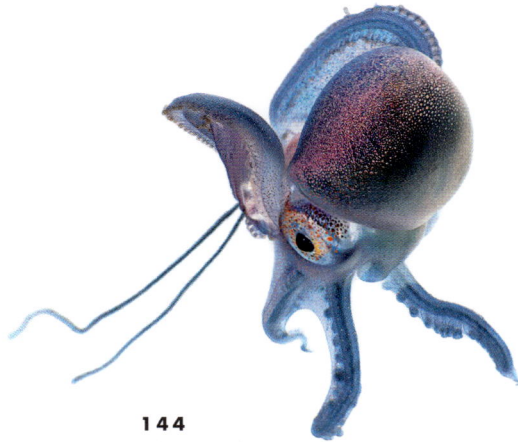

INTRODUCTION

INTRODUCTION

"Octopuses Punch Fish, Sometimes for No Apparent Reason"

"Do You Have More Self-Control Than a Cuttlefish?"

"The First-Ever Footage of a Bizarre Squid Has Scientists Freaking Out"

"Why Are These Octopuses Hurling Shells at Each Other?"

"Did a Cuttlefish Write This?"

"Squid Go into Space—for the Sake of Humanity"

→ An Atlantic White-Spotted Octopus nestles in a crevice. Like many cephalopods, it can cram into tiny spaces and squeeze through narrow openings.

This sampling of headlines from 2020 and 2021 illustrate the wonder, mystery, and downright weirdness of cephalopods—the group of animals that includes octopuses, squid, and cuttlefish. But headlines can be notorious clickbait. What's the truth behind the hype?

Every group of animals on Earth is unique in some way, but (with apologies to George Orwell) cephalopods seem to be more unique than others. They are invertebrates, animals without backbones, which places them in the same category as worms, bugs, clams, and snails. In fact, their closest relatives are clams and snails, animals with simple brains, limited vision, and the basic behavior of hiding inside a shell to escape danger. How can they be related to octopuses and squid, animals with complex brains, eyesight to rival our own, and behaviors that range from fish-punching to deliberate self-control—not to mention their breathtaking camouflage abilities?

The apparent contradiction of what hierarchically minded humans might call a "lower" animal displaying such "advanced" traits is part of why cephalopods have captured our imaginations since ancient times. But these animals have also grabbed us for a variety of other reasons. Giants of any kind fascinate us, and cephalopods include the world's largest invertebrates, with Giant Squid (*Architeuthis dux*, pages 204–205) that stretch 43 ft (13 m) long. Natural beauty entrances us, and cephalopods are some of the most beautiful invertebrates, from the shells of nautiluses and argonauts to the flashy skin of cuttlefish and octopuses.

They are also literally grabby. Cephalopods are the only animals out there with both arms and suction cups, offering us the perfect metaphor for anybody who acts exceptionally grasping and greedy. We humans do love a good metaphor. Political cartoons have used octopuses again and again to represent entities from corporations to communists, and when *Rolling Stone* in 2009 compared the investment bank Goldman Sachs to a Vampire Squid with a "blood funnel," the metaphor was compelling enough to be quoted and referenced for years. (Even after 2012, when scientists discovered that real Vampire Squid (*Vampyroteuthis infernalis*), rather than using funnels to suck blood, uses delicate filaments to collect detritus.)

We tell myths about cephalopods, like the Kraken of Scandinavia and Kanaloa of Hawaii. We make cephalopod art, from Minoan octopus vases to gyotaku or fish-printing with squid ink. In 2021, the town of Noto, Japan, invested pandemic-relief grant money in building a giant squid statue to boost tourism.

Of course, humans have also been interested in learning the truth of what cephalopods really are and what they can really do for a long time. Scientists study cephalopods for all the same reasons outlined above—plus a few more. In addition to color-changing skin, many cephalopods have bioluminescence, and the glow of bobtail squid in particular has become central to the entire study of symbiosis. These small cephalopods produce light thanks to a fascinating and complex relationship with symbiotic bacteria that live in a dedicated light organ. (These are the squid that were sent into space "for the sake of humanity," since we've learned that we, too, depend on our symbionts, and it would be good to know how such a dependency handles zero gravity.)

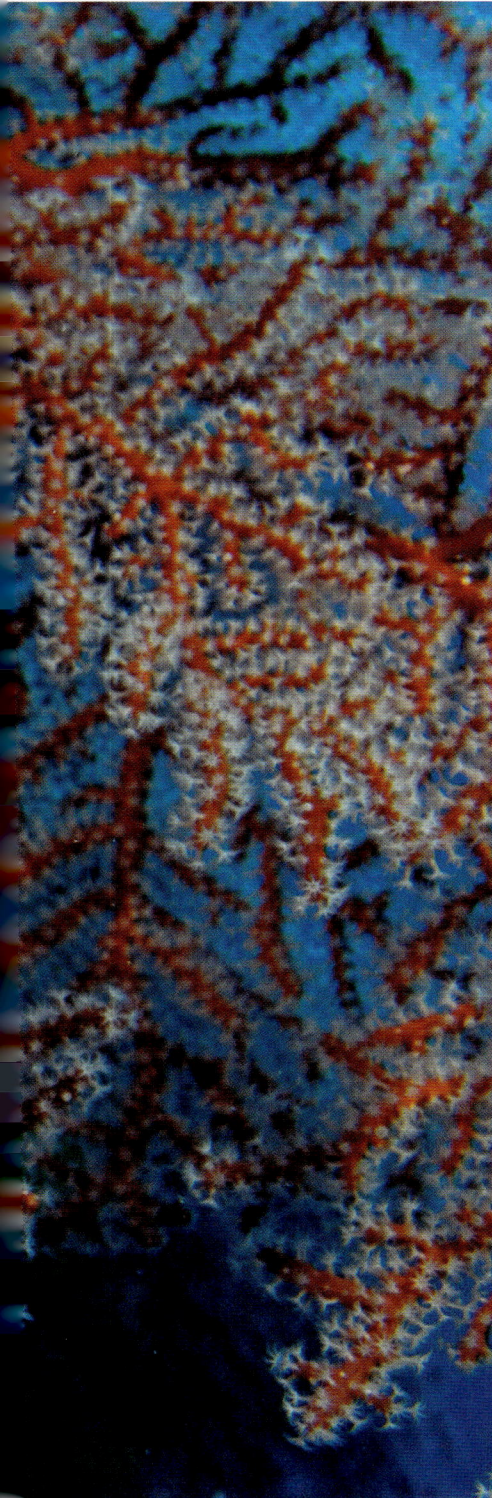

Scientists also seek to learn about cephalopods because they're a significant food source. Octopuses and squid have long been fished globally for human sustenance. They are high in protein, low in bones, and often abundant. These are also reasons that many other species love to eat them, including whales, seals, otters, fish, sharks, and seabirds. Sperm whales and some albatross species live primarily on squid. Ecological research illuminates the roles played by cephalopods in their natural environment, and can help us ensure that we don't eat other animals out of their livelihood—or drive cephalopods themselves to endangerment or extinction.

Then there are species caught not for their meat but for their beauty, such as Pearly Nautiluses (*Nautilus pompilius*) and Mimic Octopuses (*Thaumoctopus mimicus*). Nautiluses are killed so their shells can be sold to collectors, and while Mimic Octopuses are captured alive to be sold in the aquarium trade, many animals are lost along the way.

Photographs and video, rather than physical specimens, are the best way to enjoy most beautiful cephalopods. Every day, more incredible images of cephalopods are captured by professional and amateur scientists and photographers. I am immensely grateful to all these people for their tireless work illuminating the incredible world of cephalopods.

As a final aside before diving into this world, I'm contractually obligated to take a stand on plurals. Cephalopod/cephalopods is easy, while squid and cuttlefish, like sheep and fish, tend to be their own plurals. Nautilus usually becomes nautiluses, but what everyone really wants to know is: what about octopus? "Octopi" is the Latin approach, "octopodes" the Greek. Personally, I think you can't go wrong with the standard English plural "octopuses."

The first chapter of this book is an overview of cephalopod biology, while each subsequent chapter explores a major cephalopod habitat, then presents a selection of representative cephalopod species. A range map accompanies each species profile, and the available data for these maps varies widely. Some distributions consist of a few locations where the species has been observed, others comprise a large area the species is surmised to occupy based on their environmental tolerances and behaviors.

← Although it doesn't look much like an octopus or squid, the Pearly Nautilus is a true cephalopod, with its "feet" (tentacles) attached directly to its head. Instead of hiding in cracks or crevices, the nautilus can retreat into its hard shell.

WHAT IS A
CEPHALOPOD?

Introducing cephalopods

Octopuses, along with their squid, cuttlefish, and nautilus relatives, together make up a group known as cephalopods, and they are well worth getting to know better. Not so long ago, the span of a human lifetime or less, most people thought of octopuses as unpleasant or downright malevolent, if they ever thought of them at all. Political cartoonists and journalists used the octopus or squid as a metaphor for any entity they saw as grasping and greedy, from the Soviet Union to Goldman Sachs bank.

← Octopuses tend to be nocturnal, but the Day Octopus bucks this trend. Actively hunting when the environment is well lit means it must be extremely skilled at camouflage, stealth, and escape.

Marine scientists and science communicators have long struggled against this negative public perception, from marine explorer Jacques Cousteau's 1973 book *Octopus and Squid: The Soft Intelligence* to filmmaker Craig Foster's 2020 documentary *My Octopus Teacher*. The tide of humanity's attitude may finally be turning, as octopuses are now "having a moment" in the spotlight of our admiration—an observation by Sy Montgomery, author of *The Soul of an Octopus* (2015).

THE CEPHALOPOD BODY

The name "cephalopod" means "head foot," and describes the anatomy of these curious creatures. If you removed all your limbs and then reattached them to your head instead of to your torso (specifically, if you attached them to your lips) then your body would be arranged like that of a cephalopod. Obviously, you would still have a torso, containing your stomach and kidneys and lungs and heart, but it wouldn't have any arms or legs attached to it. A cephalopod's "torso" equivalent is called a mantle, containing their stomach and kidneys and gills and hearts (three of them—more on that later).

The bag-like mantle of an octopus is sometimes incorrectly called its head, but cephalopod heads are really just like ours, with a brain, two eyes, and a mouth like the beak of a bird. (The lack of teeth and the fact that the brain encircles the esophagus like a doughnut are two ways in which their heads are not *exactly* like ours.) Limbs sprout in an "arm crown" all around the mouth, like the limbs you hypothetically attached to your lips.

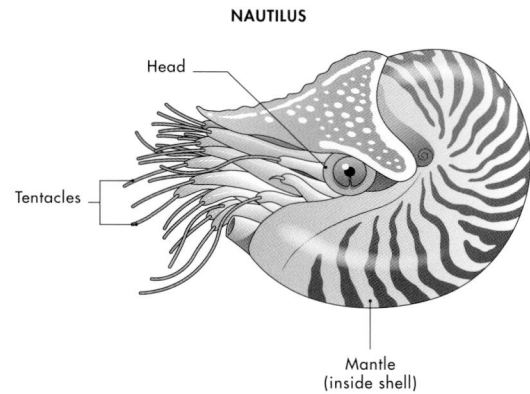

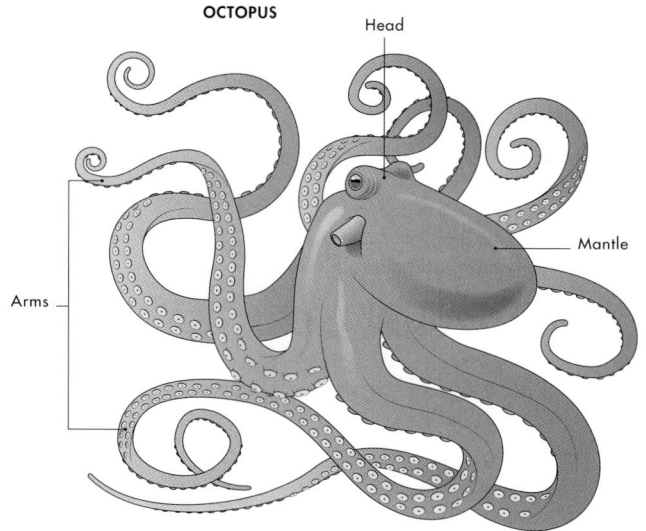

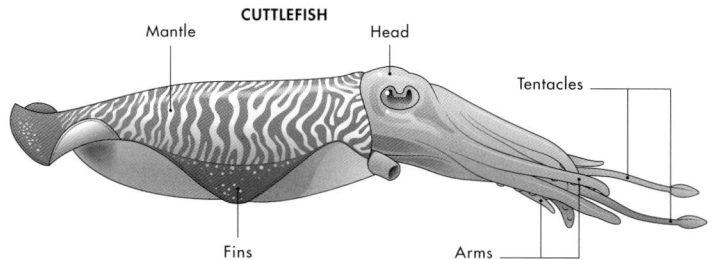

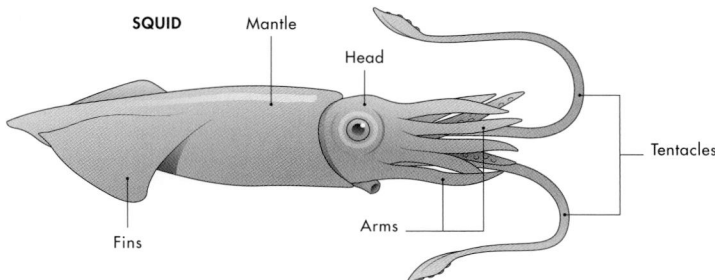

OCTOPUS

Head

Mantle

Arms

NAUTILUS

Head

Tentacles

Mantle
(inside shell)

CUTTLEFISH

Mantle

Head

Tentacles

Fins

Arms

SQUID

Mantle

Head

Tentacles

Fins

Arms

Cephalopod anatomy

The diversity of cephalopod forms all share "feet" or appendages connected to the head, which is connected to the mantle.

ARMS VS. TENTACLES

Although their name includes the Greek word for "foot," in English we refer to a cephalopod's appendages as arms or tentacles. Technically, the two terms aren't synonymous. "Arm" describes a limb lined with suckers from base to tip, like the eight arms of an octopus. Squid and cuttlefish have eight similar arms, and in addition they have two long tentacles—stretchy elastic appendages that can be tucked into pockets or shot out at high speed to capture prey. These tentacles usually bear suckers only at their tips. Cephalopod suckers are far more impressive than artificial suction cups. We'll learn how in the next chapter (see page 59).

Now that we've straightened out the definitions of arms and tentacles, we must contend with the reality that biology never fits in tidy definitions. Nautiluses, the only living cephalopods that grow shells to which their bodies are attached, have dozens of tentacles without any suckers. There are also quite a few squid species that bear hooks instead of suckers on their appendages (see pages 197–199).

Evolutionarily speaking, all cephalopod appendages evolved from a single body part: the molluscan foot, the same body part that a snail uses to crawl. In the early days of cephalopods, this foot grew into ten equal limbs. Several different lines of evidence suggest that the ancestors of modern cephalopods all had ten arms. Over geological eras, one pair of squid and cuttlefish arms evolved into tentacles, a different pair of octopus arms disappeared, and all ten nautilus arms proliferated.

→ This Common Octopus shows the two rows of sucker cups typical of most octopuses, along with the suckers' gradual reduction in size toward the arm tips.

↓ This cuttlefish demonstrates the use of arms for holding prey (in this case, a fish) that is too large to be swallowed in one bite.

Five hundred million years of fossils

Cephalopods have been on the planet twice as long as dinosaurs (including modern birds in the dinosaur category). In fact, cephalopods could be called "the new dinosaurs," because their fossils have all the same fascinating features. Ancient cephalopods reached gigantic sizes—probably the first animals on the planet to grow significantly larger than a human. They had bizarre shapes—corkscrews, paper-clip curves, knots, cones, disks, and spirals. Ancient cephalopods were fierce predators—carnivores that devoured trilobites, fish, and each other.

THE EARLIEST CEPHALOPOD FOSSILS

Of course, not all cephalopods were all of these things. (Not all dinosaurs were terrifying tyrannosaurs or seismic sauropods.) Plenty of ancient cephalopods were tiny, including the earliest known cephalopod fossils from the Cambrian Period (540–485 million years ago). Until 2021, the oldest cephalopod fossil ever found was just under 500 million years old: *Plectronoceras cambria*, a tiny snail-like creature that could crawl across the palm of your hand. The similarity is no coincidence; cephalopods and snails evolved from the same ancestors.

↘ The evolutionary history of cephalopods begins at the dawn of modern animals, with new groups arising and diversifying all the way through the present day.

The large group of animals we now call mollusks contains not only cephalopods and snails, but also slugs, limpets, clams, and more obscure animals such as chitons and tusk shells. All these diverse mollusks have a soft body (the word "mollusk" comes from the Latin word for "soft") and the ability to produce a protective shell (although many molluscan lineages, as we'll see in cephalopods, have either modified or lost their shells).

Mollusks, along with sea stars and trilobites, evolved rapidly into a huge array of species during what's known as the Cambrian Explosion (541–521 million years ago, although scientists disagree about the exact range). Before, animal life on Earth had been limited to the relatively simple shapes of sponges and jellies. Ecosystems had been likewise simple, with minimal

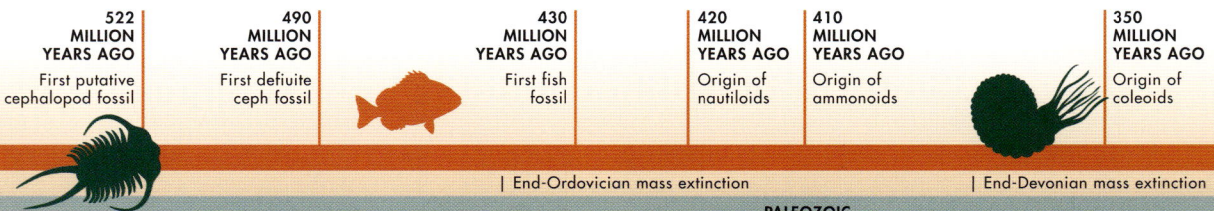

522 MILLION YEARS AGO	490 MILLION YEARS AGO	430 MILLION YEARS AGO	420 MILLION YEARS AGO	410 MILLION YEARS AGO	350 MILLION YEARS AGO
First putative cephalopod fossil	First definite ceph fossil	First fish fossil	Origin of nautiloids	Origin of ammonoids	Origin of coleoids

| End-Ordovician mass extinction | End-Devonian mass extinction

PALEOZOIC

Timeline not to scale

interactions between these animals. Then a combination of environmental changes and evolutionary feedback loops kicked in, with new animals creating niches for even more new animals to venture into.

Because *Plectronoceras* appeared near the end of the Cambrian Explosion, paleontologists initially thought that cephalopods evolved late in the diversification of mollusks. However, in 2021, fossil cephalopod shells from Newfoundland, Canada, were reported to be 30 million years older than *Plectronoceras*. (For context, 30 million years before today, there was not a single hominid or other ape on Earth.) If these fossils were indeed cephalopods—definitive identification is still underway—then cephalopods evolved alongside the earliest trilobites.

IDENTIFYING FOSSILS

How do we identify cephalopod fossils? Unfortunately, soft body parts such as arms or tentacles are rarely preserved. Instead, cephalopods present us with a distinctive fossil shell. Unlike their fellow mollusks, cephalopods evolved the ability to seal off chambers

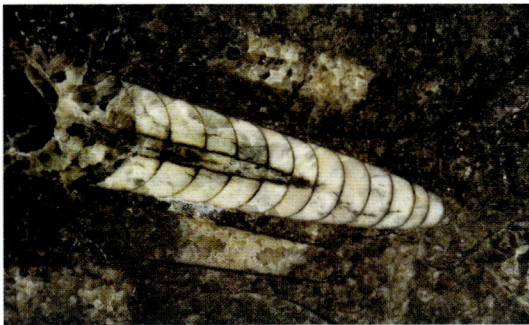

↘ Many of the most common fossils found by amateurs and professionals alike once belonged to cephalopods, since these animals were abundant and grew hard shells that fossilized well.

↙ This fossil *Orthoceras* (Greek for "straight horn") was one of the early cephalopods that grew long, straight external shells, before the evolution of groups with coiled or internal shells.

inside their shells and then fill these chambers with gas, lifting themselves off the seafloor like a dirigible. They entered a new niche that no other animal had yet explored—that of a large, swimming predator. Eventually, this niche would diversify into the entire open ocean ecosystem.

The buoyancy of the gas offset the weight of the shell itself and allowed cephalopods to evolve from tiny *Plectronoceras* to massive *Endoceras*, an animal with a long, straight shell estimated to reach nearly 20 ft (6 m), with unconfirmed reports of even longer fossils. If it had arms, which is likely, its full length could have been even greater. These giant ancient cephalopods were the first "monarchs of the sea," as described by Jacques Cousteau (1910–1997).

And then fish happened.

300 MILLION YEARS AGO
Octopus and squid ancestors diverge

228 MILLION YEARS AGO
First dinosaur fossil

100 MILLION YEARS AGO
Palaeoctopus (an early octopus with fins)

65 MILLION YEARS AGO
First primate fossil

End-Permian mass extinction | | End-Triassic mass extinction | End-Cretaceous mass extinction |

MESOZOIC | CENOZOIC

Convergent evolution with fish

The first vertebrates evolved in the Cambrian, but they took longer than cephalopods to hit their stride. Eventually, simple wormlike shapes with backbones evolved into efficient, streamlined swimmers with powerful jaws. At this point, the long, straight external shell of cephalopods became a liability. It made cephalopods too slow to compete with predatory fish for food, and they could be caught and broken by fish jaws.

EVOLVING JAWS AND SHELL SHAPES

We don't know exactly when cephalopods evolved jaws of their own, but it seems to have happened in concert with the origin of vertebrate jaws in the Silurian Period, approximately 440–420 million years ago. This is convergent evolution: when unrelated organisms evolve similar features in response to similar environments.

Cephalopods didn't have bones to work with, so their jaws were grown from hard plates of a material called chitin, shaped like a bird's beak long before birds were even a twinkle in the eye of *Archaeopteryx*. This beak houses a rasping tongue called a radula, a molluscan inheritance tough enough to drill holes in shells. As for their shells, cephalopods evolved in two different

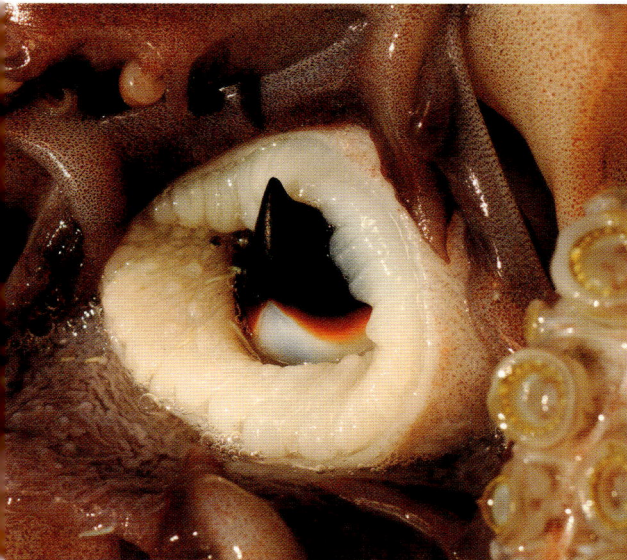

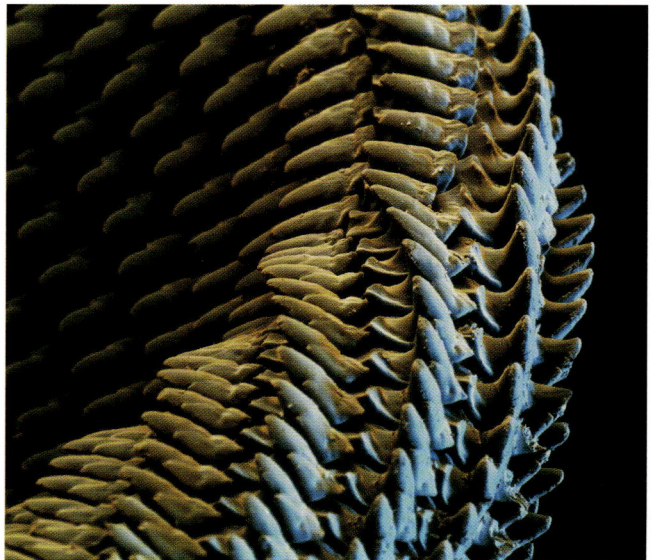

directions. Some began to grow coiled shells, which gave greater maneuverability while swimming, and were harder for fish to grasp. The first group of coiled cephalopods, the nautiloids, evolved near the end of the Silurian and remain with us today in the form of the modern chambered nautilus. A second group of coiled cephalopods, the ammonites, evolved in the early Devonian, a period that lasted from about 420 to 360 million years ago. Ammonites became so abundant that they're now some of the most common fossils for collectors to find, and so diverse that paleontologists use them to identify rocks from specific times and places. During the heyday of dinosaurs, the sea was full of ammonites that had evolved beyond the simple coil into peculiar shapes. At the end of the Cretaceous Period (66 million years ago), ammonites became extinct along with all the dinosaurs that weren't birds, but their nautilus cousins survived to share the planet with us today. That cephalopods made it through this most famous mass extinction is merely one example of their staying power. Over evolutionary history, the group has survived at least five—perhaps more—major major extinction events.

HIDDEN SHELLS AND JET PROPULSION

Meanwhile, under threat of fish, other cephalopods evolved into a group called the "coleoids." This name comes from the Ancient Greek word for "sheath." As a sword is hidden inside a sheath, so a shell is hidden inside a coleoid. Instead of packing their soft body into a hard shell, they wrap their hard shell in a soft body, turning it into a kind of internal skeleton—not unlike vertebrates, which wrap their bones in layers of muscle and skin. This shell internalization moved coleoid cephalopods much further down the path of convergent evolution with vertebrates. With their skin on the outside, coleoids proceeded to evolve streamlined fishlike bodies for efficient swimming.

Due to their evolutionary history, however, coleoids were locked into a different type of locomotion than fish. While fish swish their tails back and forth to make momentum, cephalopods swim by jet propulsion: sucking in a mantle full of seawater, then squirting it out their siphon. This is less efficient than the fish technique, because only the "exhale" part of the cycle propels the animal forward. The "inhale" part is lost time. For a fish, each swish of the tail provides propulsion.

However, cephalopods have other adaptations to enhance their movement. Many of them evolved fins to supplement, support, and in some cases nearly replace their jet propulsion. And they all have three hearts—one for each of their two gills, pumping deoxygenated blood to pick up fresh oxygen, and one for the rest of the body, pumping oxygenated blood to all the other organs.

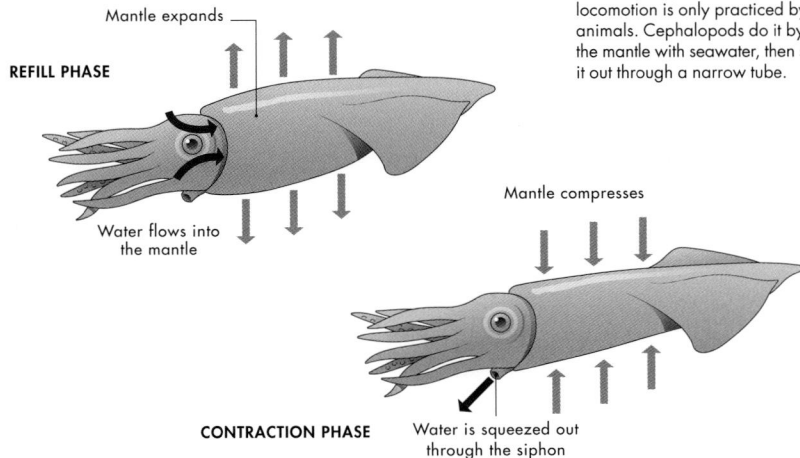

Jet propulsion
This extremely rare form of natural locomotion is only practiced by a few animals. Cephalopods do it by filling the mantle with seawater, then squeezing it out through a narrow tube.

Mantle expands

REFILL PHASE

Water flows into the mantle

Mantle compresses

CONTRACTION PHASE

Water is squeezed out through the siphon

← Seen through the technology of scanning electron microscopy, the cephalopod radula is a beautiful and complex structure. The "teeth" on this "tongue" can be used to drill through shells or draw food down the throat.

←← The beak of a Humboldt Squid protrudes from its muscular buccal mass. The hardness of the beak gradually decreases from the sharp tip to the soft base, a feature that may help materials scientists develop smoother transitions for prosthetics.

VISION

Moving and maneuvering with speed becomes a lot more useful if you can sense your environment. Thus, vertebrates and coleoids both evolved exquisite vision. In one of the world's best examples of convergent evolution, both kinds of eye focus light through a lens onto a retina filled with photoreceptors, while the different evolutionary trajectories behind each eye are obvious. Our vertebrate eyes are wired up backward, in comparison to cephalopod eyes. In humans and fish alike, nerves come out from the photoreceptors pointing toward the front of the eye, then have to make a U-turn to get back to the brain. This creates a blind spot in our vision (our brains compensate for this by interpolating the missing information). In squid and octopuses, nerves travel from behind the photoreceptors directly to the brain, leaving them with no blind spot.

The similarity between coleoid eyes and vertebrate eyes has been studied for years. In 2020, researchers discovered something else our vision has in common: stereopsis, the way we see in 3D. We humans place our eyes side by side on the front of our head, so they see the same view from slightly different angles—as you can test by closing first one eye, then the other. Our brain compares these two views to estimate distance. Special 3D glasses can trick us into perceiving depth on a flat page or a flat screen by presenting a slightly different image to each eye.

← The pupil of this bobtail squid's eye is open wide to admit plenty of light, but can be quickly constricted in a too-bright environment or to make the eye less visible to predators.

Cephalopod eyes, however, are placed on opposite sides of their head, with minimal overlap in their visual fields. This means they can take in a more panoramic view of their surroundings, but it made scientists skeptical that cephalopods could use the same technique of stereopsis to measure distance. To find out, these tenacious scientists engineered a set of 3D goggles that could be attached to the heads of cuttlefish, and showed the cuttlefish 3D movies of shrimp. Since shrimp are delicious and cuttlefish don't understand the concept of screens, they shot out their tentacles to capture what looked like dinner. The length of the tentacle strike showed researchers how far away the cuttlefish thought the shrimp was. Amazingly, this evidence indicated that cephalopods do use stereopsis to gauge the distance to their prey. Furthermore, they use different neural circuitry than vertebrates to do so—yet more convergent evolution.

SEEING IN COLOR

What about color vision? Almost without exception, cephalopod eyes do not possess the special photoreceptors, called cones, that we vertebrates use to distinguish color. Could they have converged on a totally different way to perceive it? Researchers have speculated on many possibilities, from cephalopods adjusting the size and shape of their pupils to let in different wavelengths, to color-sensitive photoreceptors embedded in their skin. But no evidence or mechanism for color vision has yet been uncovered. This makes their remarkable ability to match their camouflage to their surroundings even more mind-boggling.

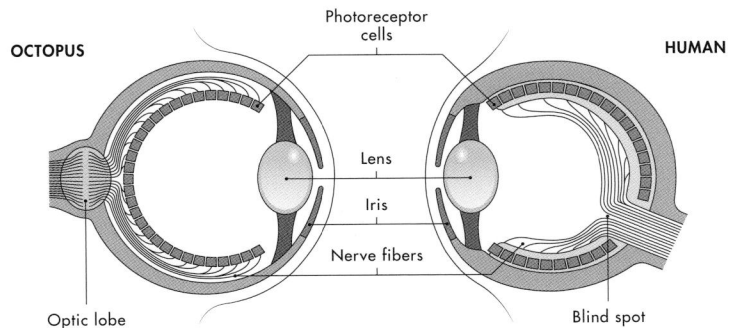

Cephalopod vs human eye
Like human eyes, cephalopod eyes have a protective cornea, an iris that expands and contracts the pupil, a light-focusing lens, and a light-sensitive retina. Unlike our eyes, cephalopods can move their lenses forward and back to change their focus.

OCTOPUS

Photoreceptor cells

HUMAN

Lens

Iris

Nerve fibers

Optic lobe

Blind spot

Meet the modern cephalopods

The classic squid body is the result of many millions of years of convergent evolution with fish. Squid are the only invertebrates that have a hope of competing with fish or mammals in a flat-out race. A squid is built for streamlined, efficient swimming. The internal coleoid shell has lost all of its calcium to become a stiff but flexible rod called a pen, or gladius. ("Gladius" means sword, carrying on with that sheath metaphor.)

The stiffness of the pen gives the muscles of the squid's mantle something to push against, so they can expand and contract with more force than those of an octopus. The pen's flexibility permits the squid's mantle to bend with the force of its swimming, but never snap. The pair of fins at the end of the mantle (sometimes called a squid's "tail") can steer and stabilize, provide extra propulsion by flapping, or wrap tightly around the mantle to make a smooth bullet shape.

SPEEDY SQUID

Even the squid's nervous system has evolved for speed. Nerve cells, or neurons, carry signals along fibers called axons. The nerves that control the muscles of the squid mantle have the largest axons in the animal kingdom, and they are capable of producing incredibly fast "escape jets." (The discovery of these giant axons kick-started the whole field of neurobiology in the first half of the twentieth century.)

↗ This rare shot of a Neon Flying Squid in the act of soaring above the ocean surface illustrates how the animal spreads its arms as well as its fins to create lift.

← The fins of this Caribbean Reef Squid help it to hover and maneuver in complex environments. If you look closely, you can see that the funnel is bent to point toward the mantle, propelling the squid in the opposite direction and making it easy to grab prey with its arms.

The open ocean squid family Ommastrephidae has even earned the name "flying squid" because of their ability to jet out of the water. However, just because the fastest cephalopods are squid doesn't mean every squid species is incredibly fast. An entirely different group of open ocean squid are called "glass squids" (Cranchiidae) due to their transparent, gelatinous bodies, and they aren't winning any races. They live a lifestyle more akin to a jellyfish.

As an aside, several groups with "squid" in their common name are not, in fact, squid. Bobtail and bottletail squids (Sepiolidae and Sepiadariidae) are more closely related to cuttlefish. The Vampire Squid (*Vampyroteuthis infernalis*) is an octopus cousin. Reef squids (*Sepioteuthis* spp.), by contrast, are true squid that look almost exactly like cuttlefish. Their paired fins run around the perimeter of their mantle like a skirt and are used to hover in place. These squid rarely jet long distances. On the inside, though, they have a proper squid pen—quite different from the internal shell of a cuttlefish.

BUOYANT BODIES

The internal shell of a cuttlefish is called a cuttlebone, with "bone" referring to the fact that it's made of calcium, like an external nautilus shell. Unlike a squid pen, the cuttlebone is relatively stiff. It has kept the buoyancy of the ancestral cephalopod shell, but instead of a series of large chambers, it contains many tiny chambers, almost like a porous sponge. With the gas in its cuttlebone, a cuttlefish hovers at perfect neutral buoyancy, needing to expend no effort to prevent itself from sinking or floating.

This limits cuttlefish to relatively shallow waters. Their cuttlebones will implode at depths below 2,000 ft (610 m). Most species prefer to stay in water of less than a few hundred feet depth. They have specialized for existence in shallow environments—hence the evolution in both cuttlefish and reef squid of the full-perimeter fin, which can undulate continuously instead of flapping like a typical squid's fins. This doesn't produce fast swimming, but it is ideal for hovering and manuevering in complex nearshore environments that have many potentially damaging obstacles, such as rocks and corals.

Squid, without any gas in their pens, rely on other techniques to stay at their desired depth. Some swim continuously. Others, such as glass squids, have evolved bodies full of ammonia, which is lighter than seawater and counterbalances the heavier parts of their bodies. An advantage of these approaches is that they can work at any depth, which gives squid access to the deepest parts of the ocean (see pages 212–247).

The cuttlefish body appears to be more limited in evolutionary potential than that of squid. While reef squids evolved lifestyles nearly identical to those of cuttlefish, there are no speedy flying cuttlefish, no transparent glass cuttlefish, no deep-sea cuttlefish.

OCTOPUS DIVERSITY

Octopuses, on the other hand, have proven every bit as adaptable as squid. Although their ancestors had ten appendages like squid and cuttlefish, they evolved away two, leaving them with only eight arms. Their close relationship to the Vampire Squid was partially confirmed by this. Despite the fearsome name, this species is not a predator, living instead on "marine snow," a term for the bits of fecal matter, dead bodies, and mucus that drift through the sea. The Vampire Squid does not have tentacles like true squid and cuttlefish but two very thin snow-collecting filaments. These evolved from the same pair of arms that was lost in octopuses.

One entire group of octopuses—cirrate octopuses (suborder Cirrina)—have fins like those of squid, and members of this group include dumbo octopuses (*Grimpoteuthis* spp.) and flapjack octopuses (*Opisthoteuthis* spp.). They live in the open ocean and the deep sea. Although they do not have a pen, they have a stiff internal structure that gives the fins something to push against. These octopuses, along with the Vampire Squid, also have very deep webs between their arms, making their arm crown into more of a bell, again almost like a jellyfish. Arm webbing is present to various degrees in other species, and it can be spread out for engulfing prey, or for contributing to the escape response from predators.

PEARLY AND CHAMBERED

Like many beautiful seashells, it was the nautilus's constructed home and not the animal inside that first captured human attention. Nautiluses produce the same nacre, or mother-of-pearl, as many sea snails and clams. However, only nautiluses grow walls separating their shell into chambers, along with small openings in these walls to allow the passage of a thin tube of flesh that controls gas exchange and buoyancy. In addition to this tube, called a siphuncle, other anatomical features include the nautilus' myriad sucker-free tentacles, its simple eye structure, and its lack of fins.

↗↗ The arms of this deep-sea octopus are almost obscured by the deep webs connecting them. When expanded, the web allows the octopus to swim like an opening and closing umbrella.

↗ What color are a Vampire Squid's eyes? The first scientists to name the animal called them red; in this photo they appear blue. In fact, the eyes themselves are clear, and reflect different colors depending on the light and environment.

← Thanks to their buoyancy, cuttlebones often wash up on shore after a cuttlefish has been killed by predators or disease. Inside a living cuttlefish, the cuttlebone is a complex structure comprising thin layered chambers with pillars for support. Liquid and gas can be exchanged between the animal's bloodstream and these chambers to regulate buoyancy.

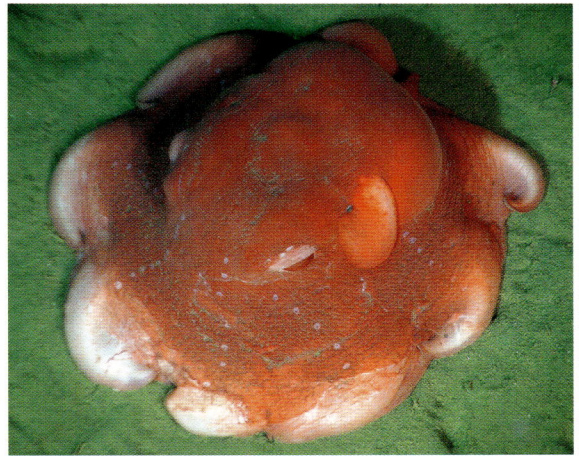

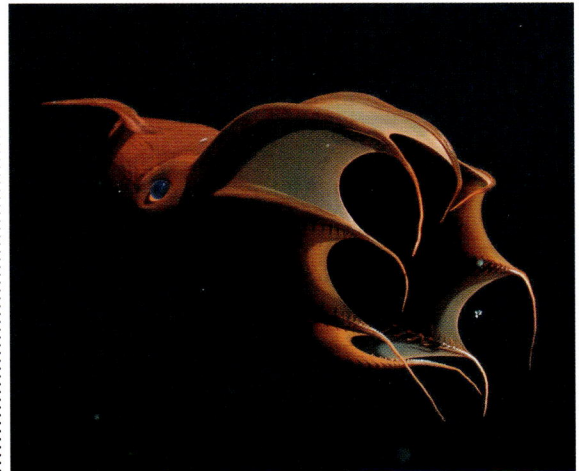

UNIQUE CREATURES

The group that comprises octopuses and the Vampire Squid (*Octopodiformes*) has been able to evolve the only known detritus-eating adult cephalopod (Vampire Squid), in addition to the only cephalopod that crawls on land (Algae Octopus), the only cephalopod that grows a shell with its arms (Argonaut), the cephalopod with the longest egg brooding time (*Graneledone boreopacifica*), and the only documented tool-using cephalopod (Coconut Octopus).

Camouflage and communication

Cephalopods are famous for their rapid color-changing abilities, which they use both to blend in and to stand out. The internet goes wild for videos of octopuses matching a background with blink-and-you'll-miss-it speed. A perennially popular aquarium exhibit is the well-named Flamboyant Cuttlefish (*Metasepia pfefferi*) showing off for visitors in bright purple and yellow.

COLOR-CHANGING TOOLS

Although cephalopods are far from the only animals that can change color (chameleons are known for it, of course, and numerous fish have also evolved the ability), their technique for doing so is unique. Other animals release hormones inside their bodies, biochemicals that disperse and alter color over a timescale of minutes to hours. But cephalopods send electrical signals from their nervous system to a network of tiny color-changing organs in their skin. They can change colors multiple times per second.

A cephalopod's skin is peppered with three kinds of "pixels": chromatophores, iridophores, and leucophores. Each chromatophore is an elastic sac of warm-colored pigment—yellow, orange, red, brown, or black. Muscles surrounding the sac can stretch it wide, turning the chromatophore "on." When these muscles relax, the sac contracts to a nearly invisible dot. Meanwhile, iridophores take care of cool colors—blues and greens and purples. They produce these colors, not with pigment, but with minuscule stacks of reflective surfaces, bouncing light at the right angle to produce the needed

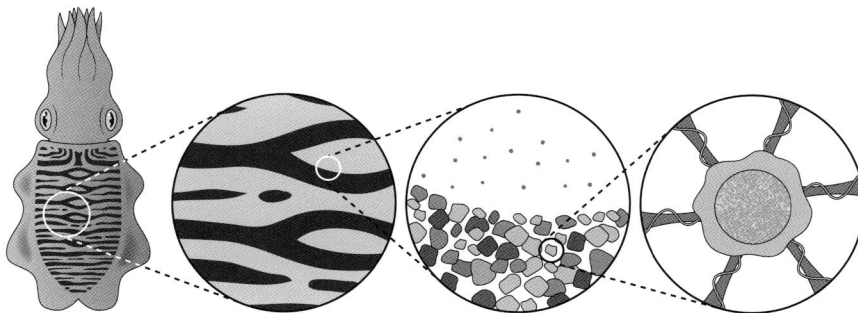

Cuttlefish

Close-up of skin pattern (e.g. stripes)

Close-up of stripe edge showing area with expanded chromatophores next to area with contracted chromatophores

Close-up of a single chromatophore

Chromatophores
The resolution of cephalopod skin is comparable to a smartphone thanks to the densely packed chromatophore organs, each of which can be expanded or contracted in less than a second.

color. Leucophores, finally, provide a background for the action of chromatophores and iridophores by reflecting ambient light. In bright sunlight, they would look white, but at various depths in the ocean, they reflect whatever available light has been filtered through the water, helping color-blind cephalopods match the surrounding hues.

These color-changing structures are complex and specialized enough to be deemed organs, like hearts or kidneys. Unlike hearts or kidneys, the density of color-changing organs in cephalopod skin varies from half a dozen to thousands per square inch. But the cephalopod capacity for camouflage doesn't end there. More muscles in their skin can be contracted to produce bumps, ridges, frills, and flaps. Thus, a smooth and featureless octopus turns abruptly into a rumpled pile of algae.

↑ An octopus blends in with its surroundings through a combination of color and texure changes. Large and small skin bumps called papillae are raised and lowered as needed, while the animal's chromatophore organs—which number over a million—are individually turned on or off to produce different colors and patterns.

↑ This cuttlefish is demonstrating a type of camouflage known as disruptive coloration. The large patches of dark and light help the animal fit in with a patchy background, as well as break up its overall shape, making it less recognizable.

↗ In 2015, scientists found that cephalopod skin contains the same light-sensitive proteins found in eyes, suggesting cephalopods may have a limited ability to "see" with their skin, to help match their camouflage to their surroundings.

COMMUNICATING WITH OTHER SPECIES

One obvious function of all these skin-changing tools is to help the animal blend in with its environment. Cephalopods strive to be unseen by both predators and prey, so they can sneak up on their own snacks without becoming one. But some species also use their abilities to be seen. The bright coloration of the Flamboyant Cuttlefish could warn predators of their toxic flesh. The vivid markings of the blue-ringed octopuses (*Hapalochlaena* spp.) advertise their venomous bite. The Mimic Octopus (*Thaumoctopus mimicus*) and the Wunderpus (*Wunderpus photogenicus*), while not particularly dangerous themselves, excel at imitating animals that are.

→   Pharoah Cuttlefish, like most other cuttlefish species, make extensive use of their color-changing abilities during competition for mates. High-contrast white and brown stripes are a common component of these displays.

↘   The eponymous markings of a blue-ringed octopus are barely visible when the animal adjusts its coloration for camouflage (left). But when it is threatened, the octopus creates a vivid warning of its venomous nature (right).

Cephalopods may even use their color-changing abilities to distract or "hypnotize" prey. Divers have observed cuttlefish displaying a pattern of rolling white and black lines as they prepare to capture a crab, and the crab appears entranced. Of course, that's our human minds reading the interaction, and further research is warranted. However, we do know for sure that cephalopods use their skin to communicate with each other.

COMMUNICATING WITH OTHER CEPHALOPODS

Cephalopods do not make sounds, so they can't talk to each other like humans or sing to each other like whales. But the combination of complex skin and excellent eyesight means they can communicate with each other visually in ways that we may never be able to decode.

A quick aside about eye-centrism: because nautilus eyes are more rudimentary, nautiluses were historically considered less intelligent than coleoids. Nautilus eyes have no lens or cornea and cannot form complex images. However, the chemical-sensing capacity of their tentacles is incredible, and recent work suggests we are just beginning to understand the complexity of their behavior. Being visual creatures ourselves, it's far easier for us to observe and track the vision-based behavior of coleoids.

The best-studied type of communication in coleoids, as it is for most animals, is courtship and competition for mates. But reproductive displays can

also overlap with other types of interaction. Octopuses may use the same colors and body shapes to intimidate each other, both when competing for mates and when competing for access to a nice den.

Having your entire body available as a canvas for communication leads to remarkable configurations. Male cuttlefish have been observed to divide their body down the middle, displaying a mating pattern to females on one side and a threat display to males on the other. Cephalopods also readily manipulate each other with misleading displays. Male and female cuttlefish display sex-specific patterns during the mating season. "Sneaker" males put on female colors to prevent other males from perceiving them as threats, then go ahead and mate with receptive females.

Although they can't see color, cephalopod eyes can perceive something ours can't: the polarization of light. If you've ever worn sunglasses with polarized lenses, you might have an inkling of this other world. Unpolarized light consists of waves vibrating in many different directions, while waves of polarized light are "organized" into a single direction. Light coming directly from the sun isn't polarized, but reflection off various surfaces, including water, polarizes it, and artificial light from screens is often polarized from the start. That's why, when you wear polarized lenses, you can tilt your head 90 degrees and change whether or not you see light bouncing off water, or being emitted from a screen. Cephalopods might be able to deliberately alter the polarization of their skin displays, giving them a channel for private communication.

But how much do cephalopods even need to communicate with each other? Is there any reason for it to go beyond: "I'm interested, let's mate"? Are they social creatures, or solitary? The answer varies tremendously by species, and we're still discovering surprises. At one extreme are octopuses that only ever approach one another to mate, and do so at arm's length because of the great risk of cannibalism. At the other extreme are reef squids (*Sepioteuthis* spp.) which live in large groups including individuals of various ages and sizes throughout their lives. Some scientists have begun referring to groups of squid as squads.

Even typically solitary octopuses have been recently shown to form temporary hunting parties, not with other members of their species, but with fish. Yes, these two long-standing rivals, whose ancestors diverged hundreds of millions of years ago and which have been competing with and preying on each other ever since, demonstrate enough brain capacity and behavioral flexibility to hunt cooperatively. Scientists have documented octopuses working with at least two different kinds of fish, groupers and goatfish, to flush prey out of hiding and take turns eating it. It's a complex social interaction, including recorded cases of octopuses punching their fish partners away from food, perhaps in retaliation for not pulling their weight in the arrangement.

In another striking discovery of cooperation, one octopus species was recently discovered to form pair bonds, with two individuals rearing multiple clutches of eggs together.

Cephalopod life cycles

Cephalopod reproduction can be a dangerous affair, with males fighting over females and females cannibalizing males. Anatomy and behavior vary from species to species, but typically males have a modifed arm called a hectocotylus, which delivers packages of sperm to the female. Males of many squid and cuttlefish species hold onto a female's body during sperm package transfer, while among octopus species, delivery tends to occur at literal arm's length, to reduce the risk of cannibalism. Yet there are always exceptions. The Larger Pacific Striped Octopus (LPSO), was reported in 2015 to mate in an intimate face-to-face position, putting those sharp beaks near each other's tender parts with no apparent risk to life and limb.

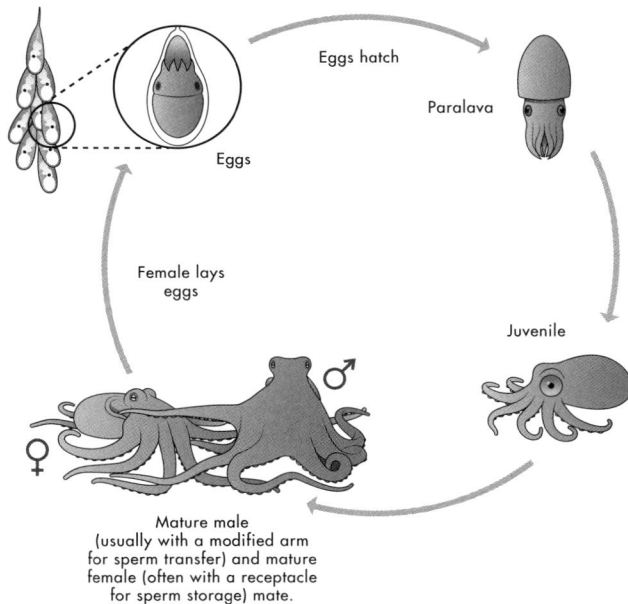

Eggs hatch

Paralava

Eggs

Female lays eggs

Juvenile

Mature male
(usually with a modified arm
for sperm transfer) and mature
female (often with a receptacle
for sperm storage) mate.

Cephalopod life cycle

This illustrates a typical cephalopod life cycle, although the many species of cephalopods exhibit many variations. Some adults brood their eggs while others do not; some eggs hatch into planktonic paralarvae and others directly into juveniles.

OCTOPUS EGG CARE

Female LPSOs can lay and care for multiple broods of eggs throughout their lifetimes. However, this is far from usual for octopuses. In 2014, a deep-sea octopus mother made headlines for brooding her eggs longer than any cephalopod, indeed, any animal, ever recorded: four and a half years. She had laid the eggs on a rock in the Monterey Canyon 4,600 ft (1,400 m) below the surface, and as scientists continued to observe the area with a remotely operated vehicle (essentially a diving robot on a leash) they never once saw her leave the eggs. She did not even attempt to capture prey items that drifted nearby. She focused on keeping the eggs aerated by blowing water over them with her siphon, and cleaning fungal growth off them with her arms.

This type of single-minded care is typical of octopus mothers, though the length of her vigil was not—it was, in fact, several times longer than the entire life span of most shallow-water octopuses. The full life cycle of many octopus species lasts only a year or less (as emotionally portrayed in the documentary *My Octopus Teacher*). But deep-sea animals of all kinds, not merely cephalopods, tend to live longer and slower lives than their sunlit relatives. In the cold water, they do everything more slowly: hunt, move, digest, grow, age.

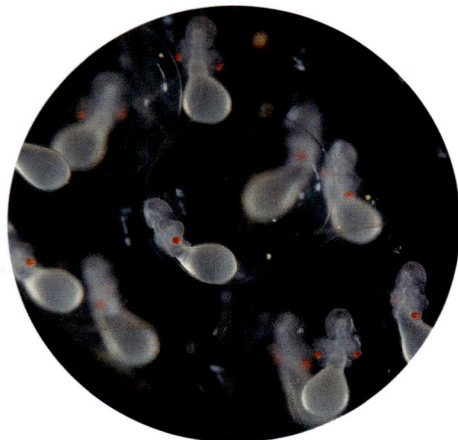

↗ These squid embryos inside their transparent eggs are attached to large yolk sacs. They'll consume the yolk as they grow and develop, until they are ready to hatch.

← A mother Giant Pacific Octopus broods her eggs, which are attached in clusters to the inside of her den. She may lay up to 100,000 tiny eggs.

Finally, the day came when the robot dived to the rocky depths of the Monterey Canyon and saw only empty egg cases. The babies had hatched, and their mother had almost certainly died. Except for LPSOs and the related *Octopus chierchiae* (pages 100–101), female octopuses stop eating after laying eggs, and their bodies slowly deteriorate. They die shortly after their babies hatch.

Male octopuses take no part in egg care, but their life spans are no longer than those of females. We don't know why. Perhaps it's simply an evolutionary side effect—there may not be any specific advantage to it, but short life spans may be genetically linked to female egg care behavior (just as male mammals have nipples, though they very rarely lactate).

SQUID EGGS

Squid and cuttlefish, like octopuses, tend to reproduce only once at the end of their lives, either in a single event or several times in quick succession. Unlike octopuses, they're liable to die promptly after laying eggs, without saving time or energy for egg care. Several squid species are famous for gathering in large reproductive aggregations, and predators and scavengers converge around them to feast on mating, laying, dead, and dying squid.

Squid eggs, by contrast, seem nearly immune to predation. This is surprising, as eggs are an extremely desirable resource, full of rich fatty yolk, and have no ability to run away. Their vulnerability is likely one reason that octopuses stay near their eggs. Squid, however, produce eggs that can protect themselves. A combination of chemicals and microbes secreted by the mother is thought to make the eggs unpalatable to predators and resistant to infection.

However, scientists were astonished to discover a squid brooding her eggs in 2005. She was a member of the deep-sea species *Gonatus onyx* (Clawed Armhook Squid), and since squid don't make dens like octopuses, she was carrying her eggs around in a large gelatinous mass. Since then, other deep-sea squid species have been found to engage in similar brooding behavior. It's likely that even more peculiar reproductive habits await discovery in the deep.

SHOTGUN REPRODUCTION

However cephalopod eggs are cared for until hatching, whether by a watchful parent or by chemicals and microbes, after hatching, the babies are often lost to predators in large numbers. That's why most octopuses and squid produce hundreds, thousands, or, in some cases, millions of eggs. The more eggs produced in a given species, the smaller each egg typically is, representing a relatively low investment of parental energy per offspring. This is the typical coleoid life-history strategy, and so far it's worked very well. They produce huge numbers of offspring with great frequency—remember, generation time is often a year or less—to withstand a vanishingly tiny success rate. The vast number of doomed young cephalopods support a great many predators of various sizes.

BABY CEPHALOPODS

In many species, these numerous hatchlings are not only much smaller than their parents, but shaped differently as well. They hatch out of their tiny eggs as paralarvae. This term was invented to describe hatchling octopuses and squid that don't look much like their parents and don't live like them either— they drift with the currents before growing large and strong enough to settle down. However, they don't undergo a metamorphosis the way true larvae like caterpillars do. Coleoids often grow from rice-grain-sized paralarvae to watermelon-sized or even human-sized adults in less than a year, rapidly assimilating energy from the bottom of the food chain and, in turn, making it available to large predators at the top.

However, a few coleoid species lay a smaller number of larger eggs. That deep-sea octopus, for example, invested four years in brooding "only" about 160 eggs. Far more babies than a mammal, far fewer than most other cephalopods. The length of brooding and the size of the eggs meant that the hatchlings emerged at a substantial size and were capable of living like adults. They were therefore not classified as paralarvae.

Nautiluses also do not have paralarvae; their life cycles are quite different from those of their coleoid cousins. Instead of living, spawning, and dying in less than a year, they mature slowly, only beginning reproduction after several years of life as a juvenile. They go on to reproduce serially, each time laying only a few large eggs. Nautilus embryos develop for several months within the egg, then hatch out with fully formed shells, tentacles, and an ability to swim and forage like adults.

← This paralarva of a Bigfin Reef Squid shows very different proportions of fins, mantle, head, and arms compared to an adult. With few chromatophores, its skin is mostly transparent, allowing us to see the highly reflective ink sac.

→ This young female Argonaut Octopus has already begun construction of the shell that she will use as an egg case when she becomes a mature adult.

Predator and prey

Even the tiniest cephalopod paralarvae were thought to hunt for their food, until recent work on flying squid (Ommastrephidae) paralarvae revealed that they do not attack prey right after hatching. Instead, they scavenge bits of dead bodies, feces, and mucus. But they soon move on to capturing the tiniest possible prey: shrimplike plankton that feed on individual algal cells. As the squid grow, they integrate that planktonic nutrition into their bodies, building themselves big enough to go after larval fish and crabs, until eventually they can target adults of these same species.

HUNTING AND SCAVENGING TECHNIQUES

Cephalopods tend to be generalist predators, ready to attack whatever prey is available. In practice, any given environment tends to present a consistent array of prey species, so cephalopods can and do specialize. Some octopus species focus on drilling into snail shells and slurping out the contents; others munch primarily on crabs. Nautiluses, for their part, go after the cast-off molts of lobsters to get the calcium they need to build their own shells. And the Vampire Squid (*Vampyroteuthis infernalis*) collects detritus, similar to the dietary habits of flying squid paralarvae, but on a larger scale.

Cephalopods use their appendages to catch food: if tentacles are present, they can be shot out in a high-speed tentacular strike, almost like a frog's tongue going after a fly. Arms can capture food too, and if there are webs between those arms, they can engulf the prey in a deadly parachute. Appendages are sometimes considered part of a cephalopod's gape, the width of a predator's mouth. Gape size determines how big a prey item the predator can go after, and a cephalopod is not limited by the relatively small size of its beak. As long as it can keep the prey trapped in its arms, then it can take bite after bite until everything is consumed. What's more, some species have venom that can dissolve biological tissue, further breaking down large prey items.

PREDATORS ON CEPHALOPODS

Over the course of their lives, cephalopods integrate energy and nutrients from a wide range of sources into a single delicious package that is, in turn, eaten by a huge range of predators—including humans. So far, we haven't managed to eat any cephalopod species to

↗ Anemones, although sedentary, are voracious predators and scavengers. This small cephalopod may have been caught alive by the anemone's stinging tentacles or it may have been collected after dying from other causes.

← As exploratory hunters, cephalopods like this octopus may encounter new potential prey items, and learn through experience whether they are worth consuming.

extinction. It would be nice to keep it that way, especially because squid are an integral part of so many other animals' diets: toothed whales, dolphins, porpoises; seals and sea lions and otters; seabirds, especially albatrosses; fish of many kinds, from sharks to rockfish to eels. All of these animals depend on an abundance of cephalopods throughout the world's seas. There's a reason that frozen squid is one of the most popular bait items.

Given the evolutionary flexibility of cephalopods, displayed over hundreds of millions of years of evolution, it's actually surprising that they're all carnivores. Why haven't any cephalopods evolved to eat seaweed? Why haven't they evolved to farm algae inside their bodies? Plenty of other "carnivorous" animals, such as sea anemones and sea slugs, have mastered that trick. It's a mystery we can hold onto as we explore all the diversity we do know about.

The rest of this book will be organized by habitat, so we can visit cephalopods in their own homes. We will explore the physical, chemical, and biological properties of each habitat before meeting the cephalopod species that reside there.

BEACHES,
TIDE POOLS, SANDFLATS,
& MUDFLATS

At the interface of land and sea

Some of the most striking human–cephalopod interactions occur when a cephalopod crawls out of the water, moving into our habitat seemingly to investigate us. One viral video in 2018 recorded an octopus at the Fitzgerald Marine Reserve in California oozing across seaweed-covered rocks, hauling the additional weight of a dead crab, which it deposited at the feet of the filming human before making its way back to the water. "What a friendly dude," the camera operator commented. Another viral video in 2021 recorded a more aggressive interaction on a beach in Western Australia, where an octopus struck a beachgoer several times with its arms, earning a reputation as the "angriest octopus" on the beach.

MOVING ONTO LAND

Do our anthropomorphic interpretations hold any water? Perhaps these octopuses were attracted to, or confused by, some stimulus we wouldn't even notice. Is there any way to determine the "real" motivations of cephalopods in these interactions, or can we only make imperfect guesses? What's certain is that very few ocean animals come onto land voluntarily. Seals and otters are exceptions, since they give birth on land, but truly marine mammals like whales and dolphins are in trouble if they get stranded on a beach. A few unusual fish can propel themselves out of water and survive for a time, and crabs can trot around on land and even, in the highly unusual case of the coconut crab, climb trees. However, the vast majority of marine life sticks to a habit of full submersion.

The space where sea meets land seems like it ought to be a barren expanse, where neither terrestrial nor marine life can thrive. However, this habitat has been around for longer than life has existed on Earth,

and many groups of animals have evolved adaptations to inhabit this liminal space. Explore any rocky tide pool or sandy beach, and you'll find a range of barnacles and clams, crabs and anemones, worms and snails. And, if you're lucky, cephalopods.

Cephalopods tend to be some of the hardest tide pool inhabitants to spot, which might be why we get so excited when we do. One reason you're less likely to see a cephalopod than a snail is that cephalopods are relatively large predators, and any ecosystem can only sustain a limited number of those. Think of a forest, and how many more rabbits and deer it contains than wolves. Another reason we rarely see cephalopods on the beach is that they tend to be night-active, while humans tend to be day-active. And finally, of course, cephalopods have those remarkable camouflage abilities (see pages 30–33).

OCTOPUSES AND HUMANS

Elusive as they are, however, the cephalopods at the interface of land and sea were the first cephalopods that we humans became acquainted with. These are the octopuses that found their way into ancient paintings and frescoes and onto pottery. People have discovered time and again that octopuses will happily climb into and inhabit human-created objects, from ceramic jugs to discarded cans and bottles. Fishing for octopuses by putting pots in the water and then hauling them back up is a long-standing tradition.

↗ Octopus vases are recognizable pottery from the Minoan civilization, which thrived on Crete and other Aegean islands 5,000 years ago.

← Many animals are adapted to life in these transient habitats, alternately connected to and disconnected from the rest of the sea.

← Although most species are nocturnal, octopuses can occasionally be spotted by daytime beachgoers.

→ The markings of a blue-ringed octopus can be hidden from view when it wants to camouflage, or be brought into bright emphasis to frighten away predators with the message: "I'm venomous, don't mess with me!"

PLINY AND THE OCTOPUS

Humans must have been curious about cephalopods for as long as we've known about them. That's our scientific impulse: to ask, what are these creatures? How do they live? Why do they act the way they do? In one of the first European works of natural history, Pliny the Elder (23–79 CE) wrote about an octopus that not only crawled out of the sea, but made its way through a fence and into a factory to partake of the ancient Roman delicacy of fermented fish guts. According to Pliny, the octopus had an arm span of over 20 ft (6 m), which could have been true or could be an example of our human tendency toward exaggeration.

Of all cephalopods, octopuses predominate at the ocean edges because they are the crawlers; most squid and cuttlefish and all of the nautiluses are swimmers. Tide pools and beaches are not great places for swimming. However, we will meet a cuttlefish, and a couple of cuttlefish cousins, that have adapted to a bottom-sitting, rock-sticking lifestyle. Octopuses, though, have really spread into this habitat. Both the tiniest and the largest octopus species occur here, illustrating that this ecosystem, despite its narrow width, has a lot of niches to fill.

Our interaction with octopuses at the rim of our terrestrial home has typically been tinged with a hint of danger, to a variable extent depending on how large and how venomous the octopuses are in that particular part of the world. All octopuses have some amount of

venom in their saliva, but in most species it's not strong enough to seriously hurt humans. The exceptions are the blue-ringed octopuses (*Hapalochlaena* spp.) in the Indo-Pacific—and these do, in fact, inahbit the intertidal. They are not aggressive, however, and would much prefer to flash their bright colors and warn us away than to bite. Most cases of bites have occurred when human behavior was particularly reckless.

WELCOME TO THE INTERTIDAL

The intertidal is an interesting place to do science. It's right there next to land—you don't have to get on a boat or in a submersible or send down a robot. And yet, it's a rough environment. Scientists and their equipment must struggle with all the same challenges as the organisms making their home there. Thus, despite their proximity to us, intertidal cephalopods have retained a great deal of mystery, and many of the species profiled in this chapter and their amazing habits were only very recently uncovered.

Let's metaphorically wet our feet with a foray into the intertidal zone—the part of the marine habitat that exists at the edge of the land. Some of us humans have lives tightly tied to the tides: fishers, oceanographers, those who live and work on coastlines. Others may have never seen a rising or falling tide. Many animals, though, depend on the tides for their very existence, because the changing height of the sea creates an entire complex ecosystem.

Tides and their impacts

Depending on where you live in the world, you might have more or less experience with tides. While very large bodies of fresh water have slight tides, an ocean is necessary for a significant tide and, even then, there's a lot of variation in tidal height from place to place. In the Bay of Fundy, Canada, the height of the sea can change 40 ft (12 m) from low to high tide. In Honolulu, Hawaii, the height rarely changes more than 1½ ft (0.5 m).

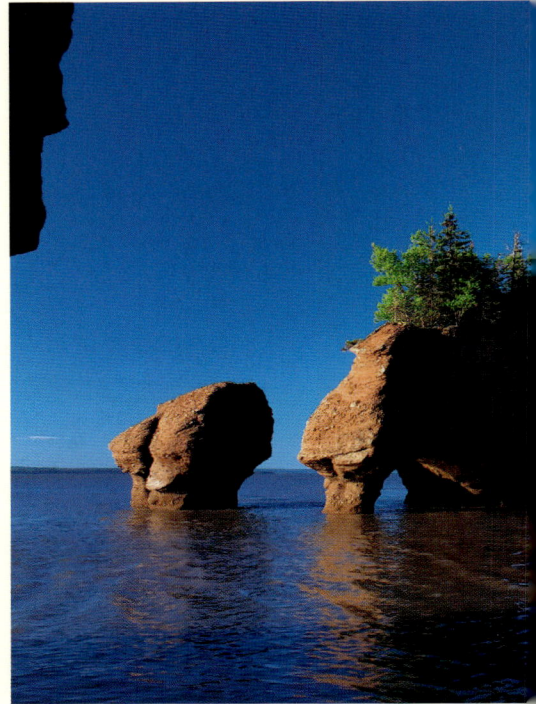

GRAVITATIONAL PULL

Tides are created by the gravitational pull of the moon, which, despite its seemingly mystical connection to the sea, actually pulls just the same on all of planet Earth: rock and soil and water alike. However, the solid parts of Earth are not easily deformed. The vast oceans, by contrast, can swell and slosh and sink.

Gravity gets stronger when masses get closer. So, the water on the side of Earth closest to the moon experiences the strongest gravitational pull, and bulges out toward the moon. Meanwhile, Earth itself is pulled slightly toward the moon, though not as much as the water closest to the moon. And the water on the far side of Earth, farthest from the moon, is also being pulled toward the moon—but less strongly than Earth itself is being pulled. So, this water on the far side also forms a bulge, since its total movement toward the moon is less than the total movement of Earth toward the moon. These two bulges of water on opposite sides of Earth

↗ If you make a large wave in a swimming pool, it can travel to the other side in less than a minute. But a wave takes hours to cross the large Bay of Fundy— almost exactly as many hours as the span between low and hide tide. This natural resonance amplifies the tidal height.

are the high tides. They always remain in the same orientation relative to the moon, but since Earth itself is spinning, different parts of our planet rotate through and experience each day's two high (and two low) tides.

TIDAL VARIATION

If that were all there were to it, tides would be the same all over Earth. But the planet is not covered with a single uniform ocean. It is covered with irregular ocean basins, hemmed in by asymmetrical continental shapes. The seas are constrained by both gravity and the physical limits of their space on the planet. When a lot of water is squeezed

into a smaller area, the tidal range (change in sea level from high to low tide) is greater. When the water has more room to move and circulate, the tidal range is less.

To add to the complexity, tides do not occur at the same time every day, but shift by about 50 minutes. This shift is caused by the "lunar day" being longer than the "solar day," because Earth keeps spinning as the moon travels around it, and the moon is always falling behind.

The height of the tide changes over the course of the year, as well. That's because the sun also exerts a gravitational force on Earth and its water. It's not nearly as strong as the gravitational force of the moon, because the sun is so much farther away, but it does play a role. When sun and moon are in alignment, both pulling in the same direction, we experience spring tides—the highest and lowest tides of the year. When they are working at odds, the moon pulling in one direction and the sun at a 90-degree angle, we experience neap tides—the most unremarkable, middling tides of the year.

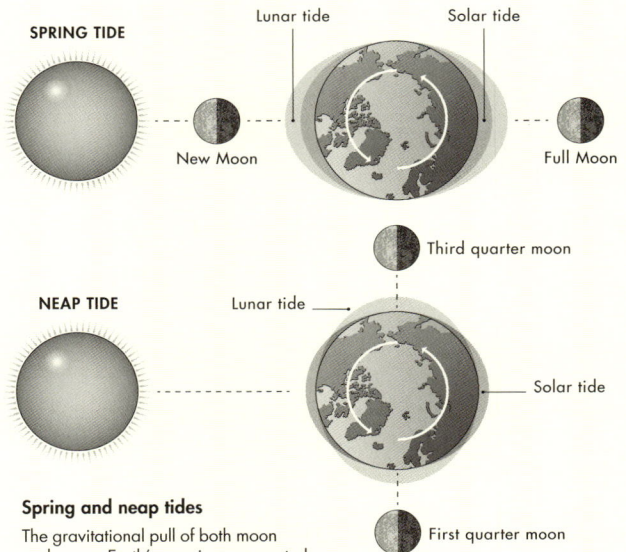

SPRING TIDE — Lunar tide — Solar tide — New Moon — Full Moon

Third quarter moon

NEAP TIDE — Lunar tide — Solar tide — First quarter moon

Spring and neap tides

The gravitational pull of both moon and sun on Earth's seas is exaggerated to illustrate how they can reinforce each other to produce the twice-monthly spring tides, or work in opposition to produce the twice-monthly neap tides.

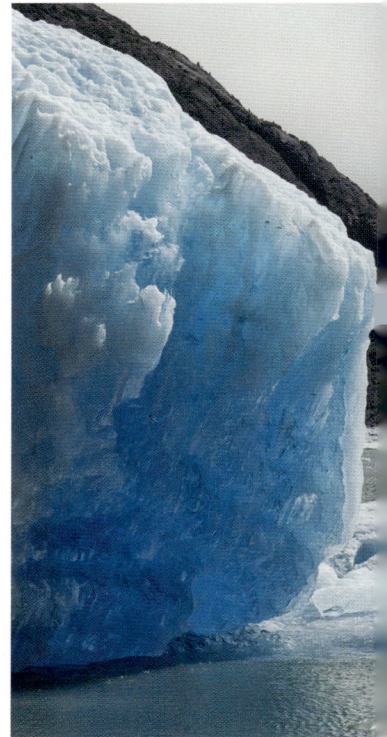

INTERTIDAL HABITATS

Every space on the coast between the highest
high-water mark and the lowest low-water mark,
is considered the intertidal. The animals that live here
have evolved to cope with an existence that not only
encompasses land and water, but also contains some
of the fiercest physical forces anywhere on the planet:
regular, daily, pounding waves, and often extreme heat
as well. If you've ever touched the water in a high tide
pool on a sunny day, you may have noticed that it can
reach bathtub temperatures.

Within the intertidal zone, each species has adapted
to its own niche. Some intertidal creatures thrive at the
highest limits, where they are almost never submerged
and get all their moisture from the splash of waves and
spray. Others can only survive in the lowest regions,
where they are nearly always wet and their daily heat
exposure is minimal. Still others can wander up and down.

↖ Mangrove forests are created by
remarkable trees that have adapted
to live in saltwater. The complex root
architecture of these plants keeps
them stable as tides rise and fall,
and creates an important habitat for
many animals.

↑ Where glaciers and ice shelves
meet the sea, the intertidal habitat is
dominated by ice. These areas look
barren compared to other ecosystems,
but many kinds of ice algae and
microbes thrive here.

↗ This American Oystercatcher has
ventured beyond the limits of its name
to catch a different mollusk: an
octopus. Shore birds are one of many
risks faced by intertidal cephalopods.

Intertidal habitats can be rocky, sandy, muddy, or a mixture. They can be dominated by algae or seagrass, although these plants (or "plantlike things," as algae is technically not a plant) cannot live too long exposed, and are more abundant further down toward the sea. Many intertidal animals burrow into the sand and rocks, which offer protection from heat, desiccation, and waves. Others, such as snails and mussels and barnacles, make hard homes they can seal shut when exposed and open up when submerged.

The intertidal is an area in constant flux, as waves erode rocks and carry sand. Organisms themselves create and build the habitat, from worms that dig holes in rock to mussels that grow expansive beds, making spaces for small worms and crabs—and, of course, octopuses—to hide and crawl and hunt in.

Intertidal zones can be as steep as a cliff face, or as smooth as a sandflat or a mudflat. These "flats" are especially mutable habitats, as a rising tide can cover a huge area with quite shallow water, and a retreating tide can leave it dry again. Usually, if you dig down at all, you can find a bit of water. These areas act as "sponges" that hold significant amounts of liquid even when they appear dry on the surface. This allows burrowing animals to stay wet through the change of tides.

Cephalopods can't cement themselves down like a barnacle, or grow a hard protection like a snail's shell. But they have their own transient and flexible approaches to the challenges of the intertidal, from borrowing the protection of other animals to building dens and burrows for themselves in rocks, sand, and mud.

Life in air and water

Generally speaking, camouflage works great as long as you are very still. But if you're a cephalopod, you can't stay still forever—you have to move to hunt for your food. This is where the turbulent intertidal environment comes in very handy for cephalopods. They can camouflage themselves to blend in with other moving things, because the sloshing, crashing waves make everything move.

ADAPTATIONS TO LIVE OUT OF WATER

Breaking waves create a "splash zone" that gets hit by spray even higher than the high tide line. This can dampen animals that would otherwise be at greater risk of drying out. However, a little periodic splashing isn't enough to keep most cephalopods happy. They face a problem that their mussel and snail cousins do not. When out of water, shelled mollusks can clam up (forgive the pun). They seal their soft, wet parts inside a mostly impermeable shell, and lose little moisture to the air. Coleoids, having given up their protective shell, cannot do this. Yet it is coleoids and not nautiluses that have colonized the intertidal. A nautilus would be useless out of water, as its arms are not muscular enough to carry its body weight on land.

↗ Unlike squid, octopuses cannot swim far with jet propulsion. The pressure it produces inside their mantles actually stops their hearts!

← Although octopuses fall far behind squid when it comes to swimming underwater, their ability to move on land is unparalleled among their squid and cuttlefish cousins.

↙ This fossil of a Jurassic belemnite illustrates how sturdy this group's internal shell used to be. But with their skin on the ouside, these animals would have faced the same challenges to life in both air and fresh water.

The large surface area of exposed coleoid skin has several disadvantages. One is dehydration, losing water across the permeable membrane through evaporation. Another aspect of the skin's permeability may explain the absence of any freshwater cephalopods. Cephalopod internal fluids, like those of all animals, have a salinity not too different from that of the ocean. Animals that evolved to live in fresh water have adaptations to maintain their internal salinity above that of their environment. Otherwise, simple chemistry dictates that fresh water would continuously diffuse into their cells as the molecules seek an equilibrium. Cephalopods have never adapted to this situation.

The permeability of cephalopod skin also brings an advantage: it can be used to breathe. Gas exchange can occur across any thin membrane, not only in lungs and gills, and many invertebrates rely on gas exchange across their skin to provide them with enough oxygen.

This "cutaneous respiration" likely plays a role in cephalopods venturing out of water, although the extent of its contribution remains to be studied.

Of course, cephalopods do have gills, like fish and crabs and so many other marine creatures. When animals with gills or lungs breathe, their aim is to bring oxygen into their bodies; specifically, into their bloodstreams. Both gills and lungs have evolved a large surface area to maximize the efficiency of transfer. Blood is passed over the inside of the surface, and air or water is passed over the outside, and oxygen-carrying molecules inside the blood bind the oxygen that diffuses across the membrane.

Zooming in on our lungs, you'd see that they are full of many tiny branching passages. Gills have a similarly large surface area, with lots of folded membranes to pass water over. However, when gills come out of the water, these folds stick together, drastically reducing their surface area and making them almost useless for pulling oxygen out of the air. Even though a cephalopod's gills are protected within its mantle cavity, the mantle is likely to deflate out of water, leaving the gills and other internal organs all pressed together until the animal can refill its mantle and fluff up its gills.

However, as long as the animal returns to water before it dries out or suffocates, there are no ill effects of such "deflation." And here's another difference between cephalopods and vertebrates: our oxygen-

binding proteins. Vertebrates have hemoglobin, which uses iron to bind oxygen. That's why it turns red (rust-red) when oxidized. Cephalopods have hemocyanin, which uses copper to bind oxygen, and it gives their blood a distinctive blue-green color. Hemocyanin isn't as good at binding oxygen as hemoglobin, which limits cephalopod activity in some situations, but in the case of crawling around on land, it may expand the time it takes them to use up the available oxygen by slowing down the process of binding and releasing this crucial compound.

WAVE ACTION

The intertidal area exists because of the tides. However, at any given moment, there's usually a far more noticeable force at work: wave action. Waves, whether huge breakers that surfers dream of or tiny ripples lapping at your feet, are not caused by the gravity of the moon or the sun. They are instead the product of physical agitation of the water, usually by wind, but occasionally by disturbances such as whales and boats.

You may have heard the term "tidal wave," and wonder where that fits into the tide/wave distinction. Because of this understandable confusion, "tidal wave" is falling out of use in favor of more specific terminology. When the rising or falling tide causes a visible rush of water, because it's being funneled through a narrow opening, that's a tidal bore. By contrast, a tsunami occurs when a huge amount of water is shifted—often by seismic activity, such as an earthquake or a volcano. Finally, a storm surge is a large rise in water level driven by meteorological activity.

All these water movements can contribute to the extreme conditions already present in the intertidal zone. However, they are sporadic, while regular wave action driven by wind is a continuous assault, minute by hour by day. The size of these ordinary waves is determined by two things: the fetch, or distance over which the wind has been blowing, and the angle of the shore onto which the wave is moving. As a wave moves from the open sea toward the shallower shore, its bottom drags on the ground and is slowed down, while the wind keeps pushing on its top. Eventually, this push–pull causes the top of the wave to "break."

Breaking waves

Waves are traveling patterns, not traveling water. As the pattern moves from deep to shallow water, the bottom of the wave is slowed by the seafloor and the top falls foward, like you might fall forward if your feet were slowed down by tripping on something.

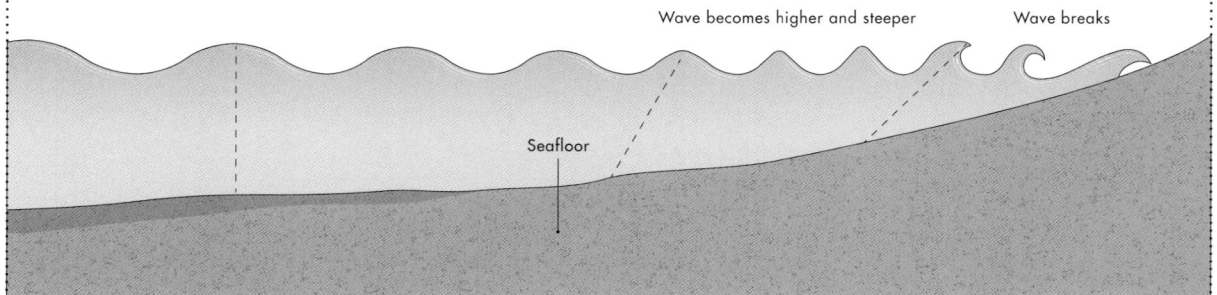

Wave becomes higher and steeper   Wave breaks

Seafloor

TERRESTRIAL PREDATORS

In addition to death by asphyxiation and dehydration, another intertidal danger is the risk of terrestrial predators—from bears and wolves to humans. All around the world, people venture into tide pools and onto sand- and mudflats to harvest seafood. Experienced fishers know how to identify an octopus den, even when to the untrained eye it looks exactly like the surrounding rocks and coral rubble. They need no sophisticated technology, not even any bait—just a sharp stick and a fast hand.

↓   A fisherman of the Dumagat people in the Philippines collects octopuses alongside other marine species. Without bones or scales, cephalopods can be an easily prepared protein for many people.

Burying and burrowing

All that wave action we've discussed leads to a lot of erosion, which is one reason we have beaches. Sand is made of tiny bits of rock broken off and worn down by year after year of pounding waves, along with pieces of shells and bones ground down in the same way.

BEACH ADAPTATIONS

Another reason for beaches is the carrying of sediment by rivers and streams. Fresh water tends to flow quickly down mountains, slowing down as it moves along less steep inclines out to the sea. Fast water can bring mud and sand and even rocks along, while slow water drops these passengers to the riverbed. Many rivers open up into deltas when they reach the coast, and this is where the sediment they've been carrying settles down.

What does this have to do with cephalopods? Well, changing your colors to match your environment works best when you have a colorful environment. In a home full of different corals and rocks, encrusted with a profusion of animals and algae, the cephalopods' ability to change their skin color and texture shines. However, on the wide expanses of sand and mud that characterize a great deal of the land–sea interface, this talent may come in less handy. Here on the flats, some spectacularly specific adaptations have evolved to cope

→ This Coconut Octopus is engaged in a combination of hiding and burying. It clings with its suckers to the inside of an empty clam shell, which is itself partially buried in the gravel and detritus of the seafloor.

← Getting sand in the eyes is an unpleasant beach experience for humans, but cephalopods that dig and burrow in sand have a transparent covering to protect their eyes.

with an environment that is physically monotonous, but temporally highly variable. Two distinct adaptations are burying and burrowing. There are many other animals that bury and burrow, but as in nearly everything else, cephalopods do it in their own ways.

DIGGING TECHNIQUES

Burying animals include humans, and you may even have participated in such activities on childhood trips to the beach. All you have to do is lie down, maybe wiggle yourself into a bit of a depression, and use hands or shovels to cover yourself with sand. Sand-dwelling octopuses and bobtail or bottletail squids will typically use a jet of water to clear a depression in the sand or mud, then settle down and use two of their arms to scoop and sweep sediment up over their bodies.

Burrowing takes more effort. You can only claim to have burrowed into the sediment if you've dug a hole deep enough to get yourself entirely below surface level. Plenty of animals do this: many different kinds of worms, clams of all sizes up to the giant geoduck, and the little sand crabs or mole crabs that many beachgoers enjoy digging up from the surf zone. However, it was only in 2015 that an octopus was discovered to be capable of such a feat (see page 66).

Covering oneself with sediment might seem to obviate any need for camouflage, but these cephalopods are usually very capable of also matching their skin to their hiding place. After all, they need to leave at least a little bit of themselves exposed to keep breathing.

How and why cephalopods make their own glue

A covering of sand sounds like temporary concealment at best, in an environment full of splashing and crashing waves. The cephalopod solution is mucus. As humans, we think of it as an unpleasant substance in our noses. In truth, mucus is formed and used throughout our bodies, and we couldn't live without it. But as important as mucus is to humans, it's even more important to mollusks.

MOLLUSCAN MUCUS

Snails and slugs on land leave mucus trails that may be familiar from gardens and sidewalks. The trail is not the point, though, it's merely a side effect of the self-lubricating foot that these mollusks use to move around.

You might think that aquatic mollusks, being in the water, wouldn't need as much mucus. In fact, they have even more uses for it. Mucus can interact with water for a huge range of applications. Thick mucus can entangle and deter predators. Mucus nets or bubbles can be used to capture food. Thin, slippery mucus can be secreted to protect the skin and wash off debris. Mucus can be mixed with pigments to create a smokescreen of ink, a habit of both cephalopods and some of their sea slug relatives. Mucus can even be a home for beneficial microbes, such as bacteria that share nutrients with their host or produce antipredator defense chemicals. This kind of

← Octopus sucker cups range from microscopic to the size of a child's hand. They regularly shed and replace their outermost layer.

Getting a grip

Tiny grooves lining the infundibulum help sucker cups grip wet, irregular surfaces. Scientists have also found that the tissue of the infundibulum is extremely soft and malleable, allowing it to mold to match the shape of whatever the octopus might cling to.

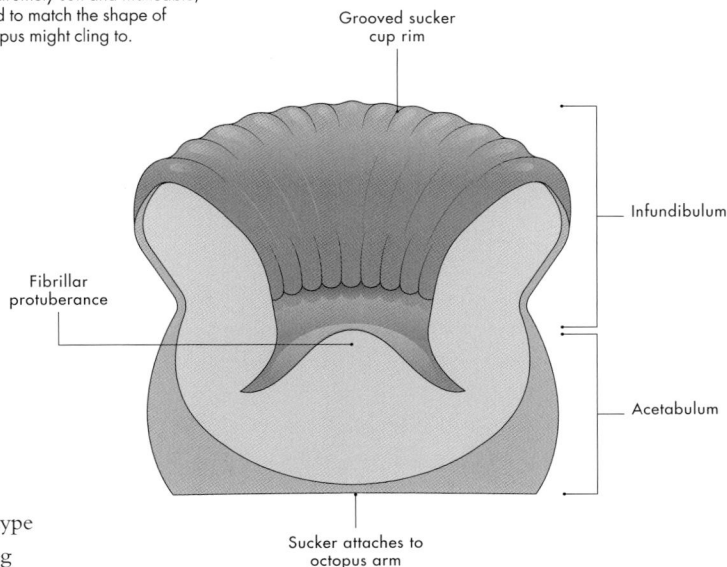

Grooved sucker cup rim

Infundibulum

Fibrillar protuberance

Acetabulum

Sucker attaches to octopus arm

interaction, with two or more organisms providing mutual aid, is a mutualism—the most well-known type of symbiosis. (Symbiosis, which simply means "living together," also includes antagonistic interactions such as parasitism. Scientists are discovering the fluidity of the distinctions between mutualism, parasitism, and commensalism, or neutral cohabitation. Environmental changes can shift a relationship from one kind to another.) And mucus can be glue.

SUCTION POWER

Cephalopods are already well known for having their own special way of sticking onto things: suction cups. Their arms are lined with hundreds of suckers, from large ones as big as a human child's hand at the thickest part of the arm of a giant octopus, all the way to nearly microscopic suckers at the slender arm tips. They create suction mechanically, by forming a seal around the rim to prevent water from entering the cavity inside the sucker, then pulling against that cavity to create a region of low pressure. This works excellently on smooth glasl, as many people have observed. However, the intertidal environment doesn't contain much smooth glass. What it does have is a lot of irregularly shaped rocks and shells and algae, with grit scattered over them.

In 2013, scientists took a closer look at octopus suckers, and found that they have a microstructural adaptation to deal with this: very small and very regular grooves along both the rim and the center of the sucker cup. The material of the sucker rim is also very soft, and this combination of roughness and softness creates a surface that can seal against nearly any other surface. One of the few substances that a cephalopod sucker will not stick to is the cephalopod itself. This fascinated the researchers who noticed it—how does a sucker recognize "self"? The answer appears to be a kind of taste. Octopus arms secrete distinctive chemicals, and their suckers contain sensory receptors that can recognize these chemicals and decide not to attach.

CEPHALOPOD GLUE

Possibly because suckers themselves are so obvious and interesting, it's taken scientists a while to study the adhesive potential of cephalopod mucus in more detail. We're just starting to get a handle on how prevalent its use is. Research published in 2007 listed only four cephalopod genera known to produce adhesive mucus; more are likely to be discovered.

Given their affinity as mollusks, cephalopods have probably been making slimes of various kinds since they first evolved. This is further supported by the fact that one of the known glue-making genera, *Nautilus*, diverged from all the others hundreds of millions of years ago. Nautiluses do not have mechanical suckers on their tentacles, and instead use mucus to make their tentacles sticky. By secreting special slime from cells in their tentacles, they can stick to prey and mates and cling to their environment as needed.

Coleoid cephalopods don't seem to make mucus with their arms; instead, they secrete it from special glands in their mantles. Several intertidal cuttlefish glue themselves down to hard substrates, such as rocks or human-created surfaces, likely to resist being buffeted by waves. The limited studies that have been done so far suggest that they use a combination of mechanical and chemical adhesion, both producing sticky slime proteins and physically contracting the center of their mantle to create a giant sucker cup.

Bobtail squids, by contrast, secrete their sticky glue all over the top of their mantles. They combine slime with burial, whisking sand over their bodies and gluing it in place for camouflage. They can release this coat of mucus plus sediment instantly, if necessary for escape. Pygmy squids (*Idiosepius* spp.) also have glue glands on the top of their mantles, though in a more specific spot close to the fins. Instead of attaching sand to themselves, they use this glue to attach themselves to the underside of seagrasses, where they sit and wait for prey to come to them. Like the bobtail squids, they can release the attachment instantly.

In both cases, the animals have two different kinds of slime-making cells. A two-part slime system can work in one of two ways. In bobtail squids, the quick release seems to be made possible by a "duo-gland" system, which creates one kind of chemical that does the sticking, and a different kind that does the unsticking. However, in pygmy squids, the two kinds of cells produce two substances that mix together to make the glue, more like a two-part epoxy. The release, then, is accomplished not by producing a third chemical, but by physically breaking the seal with the substrate.

Not all cephalopod slime turns out to be true mucus. A particular set of proteins called mucins is necessary to merit that designation, and recent protein characterization of bottletail squid slime turned up no mucins at all. It would appear that cephalopods have evolved a variety of original ways to slime it up, and we're just beginning to figure out what those chemicals are, and how they work.

← Pygmy squids are just one of many varieties of organisms that benefit from attaching themselves to seagrass. Bryozoans (also known as moss animals), hydroids (relatives of anemones), and an array of algae are some of their common companions in this habitat.

Algae Octopus
Seaweed mimic

SCIENTIFIC NAME	*Abdopus aculeatus*
FAMILY	Octopodidae
MANTLE LENGTH	2¾ in (7 cm)
TOTAL LENGTH	10 in (25 cm)
NOTABLE ANATOMY	Algal camouflage
MEMORABLE BEHAVIOR	Bipedal locomotion, "land octopus"

Memorably described by David Attenborough in a BBC Earth documentary as "the only land octopus," the Algae Octopus is certainly the cephalopod species that is most comfortable on land. It still requires seawater to live, so it's not exactly terrestrial, but it routinely crawls from pool to pool, hunting for food.

Underwater, its movement is also remarkable. Along with the Coconut Octopus (*Amphioctopus marginatus*, see pages 64–65), the Algae Octopus was one of the first two octopus species to be observed by scientists engaging in bipedal locomotion—walking on two limbs. We humans are somewhat understandably obsessed with other animals' behavior when it reminds us of our own, so this discovery made news headlines.

Apart from its humanoid appearance, bipedal walking intrigues us with the opportunity to compare its mechanisms between species. Prior to its discovery in these octopuses, the majority of known bipedal locomotion occurred in vertebrates—primates, birds, the occasional lizard. Our bones, joints, and muscles work together to create a balanced gait.

How could invertebrates, with no bones and no joints, accomplish something similar? The only one observed to do it occasionally had been the cockroach, which still has a tough jointed external skeleton.

Researcher Christine Huffard described octopus walking evocatively: "The octopus rolls along the sand as if on alternating conveyor belts." Without joints, there is no real distinction between "foot" and "leg." The end of one arm is simply laid down on the sand, then pushes off from imaginary "heel" to "toe" in a single continuous movement, while the second arm is laid down in advance of the first, preparing to repeat the pushing-off movement. The octopus's stride length is thus far less constrained than our own, perfectly embodying the quote often attributed to Abraham Lincoln that a man's legs should be "long enough to reach from his body to the ground."

The Algae Octopus does not walk on two arms in order to look like a human, but it is mimicking something else—a drifting clump of algae, as you might guess based on the species name. Wrinkling and ruffling its skin, it spreads and curls its other six arms around its body to resemble a cluster of seaweed. The two walking arms could be no more than pieces of loose algae, trailing along the bottom as the current carries the "definitely-not-an-octopus" along.

→ The scientists who first observed walking Algae Octopuses had to laugh at the visual absurdity of their gait— but they also had to analyze and appreciate the complexity of a behavior that has clearly served this species well when it comes to survival.

AMPHIOCTOPUS MARGINATUS

Coconut Octopus

Mobile homeowner

SCIENTIFIC NAME	*Amphioctopus marginatus*
FAMILY	Octopodidae
MANTLE LENGTH	3¼ in (8 cm)
TOTAL LENGTH	6 in (15 cm)
NOTABLE ANATOMY	Typical color pattern of dark branching "veins"
MEMORABLE BEHAVIOR	Bipedal locomotion and tool use

For an octopus, this species has attained a level of fame few others can aspire to. Once known simply as the Veined Octopus for its resting color pattern, it has made headlines twice already in the twenty-first century: in 2005 for its ability to walk along the sandy seafloor using only two arms, a form of bipedalism, and in 2009 for its habit of carrying around coconut halves as a mobile shelter, an activity that constitutes tool use.

Both behaviors likely evolved in response to the octopus's habitat in the tropical western Pacific, where it lives in very shallow water close to shore. Although coconut trees do not grow in the sea, the octopus lives in a habitat so close to coconut-filled islands and coasts that its sandy seafloor habitat is full of coconut shells. Waves and currents often push these shells along the sand. When the octopuses gather six of their eight arms around their head and mantle, leaving two arms free to push off the bottom, they look not unlike a bobbing coconut. To be sure, coconuts do not walk on two legs,

but neither do octopuses as a general rule, so the animal confuses its predators either way.

For their second astonishing behavior, Coconut Octopuses stack two coconut halves like bowls, then carry the stack under their arms and walk along with their arms around the outside of the shells. This isn't bipedal, but it is weird enough to get its own name, "stilt-walking," and scientists point out that it's pretty inefficient. Stilt-walking is only useful because it lets them carry protection, which they then have to stop and assemble, flipping one of the halves over to create a spherical lair.

This was considered the first evidence of tool use in an invertebrate species, where a "tool" is an object that "provides no benefit until it is used for a specific purpose." If octopuses merely hid in coconut shells when they came across them, as they hide under rocks and inside snail or clam shells, that would not really be tool use. The key innovation here is that the octopuses actually carry the coconut shells around with them, at some cost to themselves, anticipating a future use.

→ Humans use different parts of the coconut plant for food, fibers, and fuel, in cosmetics and as building materials. The usefulness of coconuts extends even beyond all that humans do with them. Resourceful Coconut Octopuses have learned to utilize the empty shells as mobile homes.

OCTOPUS KAURNA

Sand Octopus

Subterranean slimer

SCIENTIFIC NAME	*Octopus kaurna*
FAMILY	Octopodidae
MANTLE LENGTH	3⅓ in (8.5 cm)
TOTAL LENGTH	16½ in (42 cm)
NOTABLE ANATOMY	Very long arms
MEMORABLE BEHAVIOR	Subsurface burrowing

Named after the Aboriginal Kaurna people of South Australia, this octopus is distinctive in both physical appearance and habits. Its arms are extraordinarily long, and it is the only known octopus to truly burrow beneath the surface of its sandy habitat.

Although such burrowing had been suspected previously, it was first examined and studied in scientific detail in 2015. Researchers observed Sand Octopuses burrowing in the wild as well as in the laboratory, where they built aquariums in the shape of scaled-up ant farms so they could watch the process of burrow formation through clear glass.

Sand Octopuses make their burrows not by digging with their arms, but by blasting water from their siphon. These jets of water "fluidize" the sediment, in the same way that shaking a jar containing both sand and water temporarily suspends the sand particles in the water. The octopus can move right into the fluidized sediment, just like you can swim through a muddy lake even though you can't swim through mud that has settled to the bottom of the lake. The octopus first enters with its arms, keeping both mantle and funnel above the surface to suck in water and blast the burrow bigger, until there's enough fluidized sediment to fit its whole body below ground.

So far so good, but the sediment won't stay fluidized. How does the Sand Octopus keep from suffocating once it's well and truly buried? Before the sediment completely settles down, the octopus reaches two arms up to the surface to make the shape of a chimney, like a snorkel for breathing. Then it uses slime to solidify the walls of both chimney and burrow.

Cephalopods are masters of repurposing the efforts of other animals, happily occupying empty snail and clam shells. Sand Octopuses are no exception, and have been seen occupying worm burrows to save themselves the trouble of fluidizing their own sediment. Scientists don't know yet how the Sand Octopuses produce the slime, what the slime is made of, or how it's integrated into the sediment.

How the Sand Octopus burrows

1. Fluidizes sediment with a water jet
2. Slides in arms first. Repeats steps 1 and 2 until buried
3. Uses arms to shape a chimney
4. Cements burrow walls with mucus

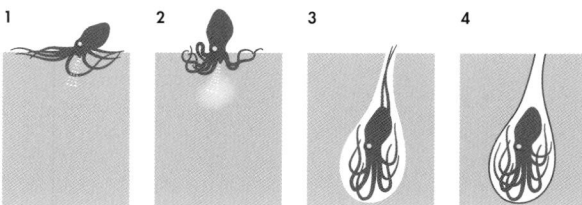

→ The Sand Octopus's habit of digging a sand burrow and securing the walls with slime is reminiscent of the garden eel, which also lives in tropical sandy habitats and makes a mucus burrow —a good example of convergent evolution in fish and cephalopods.

Atlantic Longarm Octopus

Flatfish impersonator

SCIENTIFIC NAME	*Macrotritopus defilippi*
FAMILY	Octopodidae
MANTLE LENGTH	3½ in (9 cm)
TOTAL LENGTH	12 in (30 cm)
NOTABLE ANATOMY	Very long arms, distinctive paralarva
MEMORABLE BEHAVIOR	Flatfish mimicry

This tricky octopus existed for many years in the minds of scientists as two separate species: *Octopus defilippi* and *Macrotritopus equivocus*. The former was an adult octopus that roamed shallow sandy habitats, eating crabs and burying itself in the sediment. The latter was a transparent octopus paralarva that drifted in the plankton, two of its eight arms far longer than the rest.

Eventually, scientists raised *Macrotritopus equivocus* paralarvae to adulthood and found that they metamorphosed into *Octopus defilippi*. A few years later, the connection was confirmed when adult *O. defilippi* laid eggs in captivity, which hatched into *M. equivocus* paralarvae. The two names were merged, with *Macrotritopus* becoming the new genus. (Octopus researchers tend to be eager to place *Octopus* species into new genera. For a long time, the genus *Octopus* was vastly overused as a "wastebasket taxon," a group into which every new species that didn't obviously belong anywhere else was dumped. Scientists are still clarifying the distinctive features of the genus *Octopus*, and which species don't belong there.)

Like both Algae Octopus and Coconut Octopus, the Atlantic Longarm Octopus needs to move across its shallow sandy habitat and, also like them, it prefers to look like something else when it does. It arranges its arms and mantle into a wide, flat ovoid, with the trailing tips of the arms resembling the tail of a flatfish. This flatfish mimicry was even observed in a paralarva that was reared to adulthood in captivity, although the octopus had never in its life encountered a real flounder. This suggests that the behavior is a stereotyped evolutionary adaptation, hardwired into the genome, rather than something that is learned by exposure.

One has to wonder about the advantage. Flatfish are not poisonous, so the Atlantic Longarm Octopus cannot take advantage of local predators' avoidance of known poisonous species. Flatfish are, however, full of bones, which makes them somewhat more challenging to eat than the boneless "protein bar" of an octopus. All other things being equal, there's not much better food than an octopus for a marine predator, so for an octopus to resemble almost anything else is probably a reduction in the amount of unwelcome attention it receives.

→ The unusually long arms of *Macrotritopus defilippi* give this species a lot of material to mold into shape for its characteristic flatfish mimicry.

WUNDERPUS PHOTOGENICUS

Wunderpus

Delicate beauty

SCIENTIFIC NAME	*Wunderpus photogenicus*
FAMILY	Octopodidae
MANTLE LENGTH	1 ¼ in (3 cm)
TOTAL LENGTH	8 in (20 cm)
NOTABLE ANATOMY	Dramatic fixed markings
MEMORABLE BEHAVIOR	Mimicry of venomous animals

This species was referred to as the Wunderpus before it even had a scientific name, due to its incredibly dramatic—and, unusual among cephalopods, fixed—coloration. The high-contrast brown and white bands on its arms and the splotches on its mantle are very similar to those displayed at times by the closely related Mimic Octopus (*Thaumoctopus mimicus*), but the Mimic Octopus readily changes its patterns. The Wunderpus does not.

Because it stands out so noticeably against its sandy habitat, and because it lives in shallow water relatively easy to access by recreational divers, the Wunderpus also accumulated quite a collection of celebrity photos by the time researchers gave it a scientific name. Hence the species: *photogenicus*.

If we humans can see it so well, it stands to reason that other visual predators can also spot the Wunderpus. In fact, that may be the adaptive value of the pattern. Although this species does not imitate as many other species as its cousin the Mimic Octopus, Wunderpus octopuses are known to mimic two other species with which they share their conspicuous banding pattern: the lionfish and the sea krait (a kind of sea snake). Both are extremely venomous, the lionfish injecting toxins through its spines and the sea krait through its fangs. Thus, the Wunderpus may be using a strategy called Batesian mimicry, making predators think it is dangerous simply because it resembles another dangerous animal.

Because of its beauty, both professional and hobby aquarists have attempted to keep Wunderpuses. However, these octopuses do not adapt well to captivity, losing their coloration and remaining hidden as much as possible. Even more problematic is the fact that they live only in the tropical Indo-Pacific, and we don't know how large their populations are. Collecting tropical animals for the aquarium trade always means collecting more than will be sold, because long-distance shipping is stressful and not all will survive. Any demand for these animals in captivity could be dangerous to their survival in the wild. It's much better to enjoy them in their natural habitat, either in person if you live close enough (or can travel), or by appreciating the gorgeous photos and videos made available by local divers.

→ The Wunderpus defies the octopus stereotype of hiding in plain sight by attracting attention with its bold markings—like the brightly colored butterflies on land that warn away predators with their patterns.

EUPRYMNA TASMANICA

Southern Bobtail Squid

Round shiner

SCIENTIFIC NAME	*Euprymna tasmanica*
FAMILY	Sepiolidae
MANTLE LENGTH	1½ in (4 cm)
TOTAL LENGTH	2¾ in (7 cm)
NOTABLE ANATOMY	Light organ
MEMORABLE BEHAVIOR	Burying

Bobtail squids, sometimes called dumpling squids, are widely considered the most adorable cephalopods, despite the dearth of scientific research on comparative cephalopod cuteness.

Bobtail squids in the genus *Euprymna* all have a glowing light organ on the underside of their bodies. The light organ of the Hawaiian Bobtail Squid (*E. scolopes*), a close cousin of *E. tasmanica*, has been studied in extensive detail in the laboratory. The light in the organ is produced by symbiotic bacteria that the bobtail squid acquires from its environment and carefully cultivates in internal structures called crypts.

The light is similar to a type of camouflage called countershading, common in marine animals. Countershading consists of dark coloration on top of the animal, to blend in with the dark depths to any observer looking down from above, and light coloration on the bottom, to blend in with the bright surface to any observer looking up from below.

Observers may be predators or prey, since the animal benefits from being invisible to both. Countershading is done with white or pale colors on the underside, and counterillumination—producing true light as bobtail squids do—serves a similar purpose.

Bobtail squids bury themselves in the sand during the day and emerge to hunt at night. It might seem like nocturnal activity would make counterillumination unnecessary, but the moon can be very bright. The brightness of the moon also varies over the lunar cycle, and bobtail squids have evolved the ability to modulate the brightness of their light organ to match the variable brightness at the surface.

Although its light organ is less intensively studied, *E. tasmanica* has become a popular laboratory organism in its own right. One fascinating experiment delved into personality traits of *E. tasmanica*. It found that individual bobtail squid can be reliably placed on an axis of shy to bold, or graded by degrees in traits, such as persistence. These personalities are consistent for a given animal in a given context, but not consistent across contexts—for example, one bobtail squid might always act shyly when presented with food, but boldly when presented with a threat. This is not so terribly different from humans.

Lighting up the shadows

Counterillumination is when an organism produces light on the underside of its body to match the sunlight or moonlight coming down from the surface. This counteracts the shadow that would otherwise be cast, hiding the organism from potential prey as well as predators.

→ Bobtail squid are found in every ocean and at nearly every depth. Although they are closely related to cuttlefish, their internal shells have been greatly reduced.

SEPIADARIUM AUSTRINUM

Southern Bottletail Squid

Tiny slimer

SCIENTIFIC NAME	*Sepiadarium austrinum*
FAMILY	Sepiadariidae
MANTLE LENGTH	1 ¼ in (3 cm)
TOTAL LENGTH	1 ½ in (4 cm)
NOTABLE ANATOMY	Mucus glands
MEMORABLE BEHAVIOR	Slime production

The family of bottletail squids contains only a handful of species, all of which are known for their slime production. The primary use of this slime does not appear to be as a glue or a lubricant, as so many other molluscan slimes are—no, the purpose of bottletail slime has more in common with the slime produced by hagfish (piscine slime masters of the deep sea). It is a predator deterrent.

Bottletails have mucus glands on the underside of their mantles and heads, and the slime that they produce can expand enormously in seawater—up to 20 times the volume of the squid itself. As you might imagine, a predator looking for a mouthful of cephalopod and finding itself instead with many mouthfuls of slime is not likely to target that particular cephalopod again.

→ The English names bottletail and bobtail refer to the rounded mantles of sepiadariids and sepiolids, while their Spanish name *globitos* describes their bodies as "little balloons," and their Japanese name *mimi-ika* or "ear squid" is a reference to their large, round ear-like fins.

It's possible that the slime is toxic, or at least unpalatable. One species closely related to the Southern Bottletail Squid, the Striped Pajama Squid (*Sepioloidea lineolata*), has been theorized to exhibit dramatic striped markings like those of venomous lionfish or sea kraits (or the Wunderpus mimicking them) to warn predators of its noxious slime.

Not only is this fascinating in its own right, but it makes materials scientists salivate. Biomimetics, the field of finding engineering inspiration in nature, thrives on such discoveries. Evolutionary adaptations such as bird wings, spider silk, and termite mounds have all led to creative advances in artificial construction. Similarly, proteins from marine slime have given researchers insights into tough fibers and waterproof glues. Study of the molecular structure of hagfish slime informs the development of antifouling treatments for boat hulls and other underwater materials. (Antifouling materials cause anything that tries to settle on the surface to slide right off.) Some slime fibers even rival the strength of spider silk.

After subjecting the slime of bottletail squid to laboratory tests, scientists came to the rather astonishing conclusion that it is not, in fact, a true mucus. It contains none of the characteristic proteins called "mucins" that create the sticky linkages in everything from hagfish slime to human snot. Instead, this slime is full of proteins that have never been found anywhere else—and may therefore provide clues to brand-new materials in medicine, textiles, and more.

SEAGRASS BEDS, KELP FORESTS, & ROCKY REEFS

Ecosystem engineers

Moving seaward from the intertidal zone brings us to the continental shelf, a relatively shallow stretch of sea. At less than 330 ft (100 m) depth, the continental shelf is considered part of the continental mass—land that was exposed in prehistoric times when sea level was lower. Around some continents, the shelf can be hundreds of miles wide, though it's usually much less. Where plate tectonics create a steep drop-off, the shelf can even be absent. Past any continental shelf is the continental slope, a steeper decline that leads down to the deep seafloor.

ORIGIN OF PHOTOSYNTHESIS

The inorganic shape of our planet offers a certain complexity of habitat: mountains and valleys, plains and canyons. In that regard, Earth is not too different from other rocky planets in our own and other solar systems. But how did we arrive at the stunning complexity of ecosystems created by living organisms? It started because Earth also has water, raining down on the high places, flowing over the slopes, and gathering in the low places. We don't know of any other planet with liquid water on its surface, and it seems hardly coincidental that the only known planet with liquid water is also the only known planet with life.

Life as we know it is water-dependent; life on Earth first evolved in the water. Early life-forms were aquatic single-celled microbes. Life's first big (unintentional) engineering project occurred after some of these microbes evolved the ability to create sugar from sunlight. Cyanobacteria, or blue-green algae, were the planet's first photosynthesizers— and the first producers of oxygen, a by-product of photosynthesis that can damage DNA and kill cells. They represent the one and only time that oxygen-producing photosynthesis evolved on Earth.

↗ Giant kelp can grow at the incredible rate of a foot or more per day. This rapid growth helps underwater forests recover from damage by storms and other stressors.

← An aerial view of the Great Barrier Reef in Australia shows the scale of engineering accomplished by the tiny animals that make up coral colonies. Coral reefs absorb wave energy, protecting coastlines from storm damage and flooding.

Photosynthesis first oxygenated the ocean, likely poisoning many other life-forms that had evolved in the absence of oxygen and could not tolerate it. Eventually, so much oxygen accumulated in the water that this volatile gas began to enter the atmosphere as well, leading to the Great Oxygenation Event of 2.1–2.4 billion years ago. In a world full of oxygen, life evolved techniques to not only tolerate it, but metabolize it, thrive on it, and eventually depend on it. Thus, from the beginning, life has been making new niches for other life.

FORESTS

Today, no habitat on Earth is purely inorganic. Even the ice of the poles is colonized by microbes. All life changes its environment. But some organisms do so in a more grandiose manner, and these we call ecosystem engineers. On land, trees are a prime example. Even a single tree creates an ecosystem above ground for insects and birds and climbing mammals, and below ground for more insects and worms and fungi. As for a forest, that's one of the most complex ecosystems on the planet.

Rainforests loom large in conservation programs around the world because cutting the trees extinguishes so many dependent species.

Forests grow underwater as well: kelp forests. Kelp is often thought of as an "ocean plant," but it is technically a kind of algae. Let's take a moment to explore these terms. There are three kinds of algae: red, brown, and green. (But in true biological fashion, the group is full of exceptions and additions.) Algae include single-celled organisms such as cyanobacteria, symbiotic organisms such as the algae that cooperate with fungi to create lichens, and enormous kelp forest engineers such as the giant kelp *Macrocystis* and the bull kelp *Nereocystis*. Algae that live in the sea, and are large enough to be visible to the human eye, are commonly referred to as seaweeds. Large brown algae fall into all three categories: algae, seaweed, and kelp. They do not fit in the plant category, because that group evolved from the green algae.

HOW ALGAE AND PLANTS GOT STARTED

To understand, let's go back in time again. The photosynthetic organisms we're most familiar with today, from oaks to lettuce, are a result of endosymbiosis, or one organism taking up residence inside another. Long ago, some cyanobacteria began living inside other single-celled organisms,

sharing the products of their photosynthesis in exchange for greater safety, mobility, or nutrients. Over millions of years, these cyanobacteria became more and more simplified. They lost many of their independent features and turned into a type of organelle, a specialized structure within a cell. We know them today as chloroplasts.

What's truly amazing is that, unlike the evolution of oxygen-producing photosynthesis itself, the act of endosymbiosis happened multiple times, with one cyanobacteria-containing cell getting engulfed by another, and sometimes yet another, leading to the many lineages of algae. These events are referred to as secondary and tertiary endosymbiosis, and scientists are still figuring out which types of algae resulted from which events. Land plants are one of the simplest cases, because all terrestrial plants derive from the group of green algae that evolved from a single primary endosymbiosis.

Endosymbiosis

Organelle membranes helped scientists figure out how many times endosymbiosis evolved. Each occurrence of one cell engulfing another adds two membranes to the organelle's total: its own external membrane becomes internal, and the engulfing cell enwraps it as well. Over time, however, some membranes can be lost.

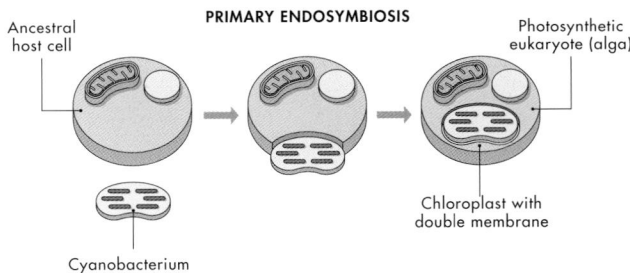

PRIMARY ENDOSYMBIOSIS

Ancestral host cell

Photosynthetic eukaryote (alga)

Cyanobacterium

Chloroplast with double membrane

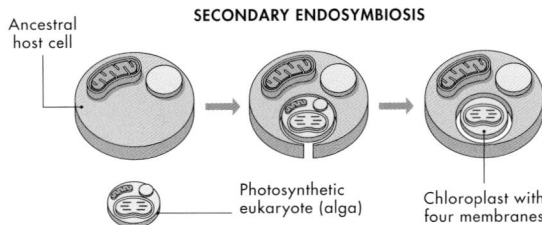

SECONDARY ENDOSYMBIOSIS

Ancestral host cell

Photosynthetic eukaryote (alga)

Chloroplast with four membranes

MARINE ECOSYSTEMS CREATED BY ALGAE AND PLANTS

Algae and plants both photosynthesize as a result of their ancestors taking on cyanobacteria as cellular roommates. That's what makes them so good at building an ecosystem—they can take energy from the sun and turn it into both structure and food for other animals. Some true plants do live in the sea: seagrasses. Seagrass beds are similar to grasslands. Seagrass can grow over large expanses and support many kinds of tiny creatures that cling to and climb around the blades, as well as substantial grazers such as dugongs and manatees. Although seagrasses are far less speciose than marine algae, they provide critical aquatic habitat.

Then there are reefs. The word "reef" tends to conjure up visions of coral, which will be covered in the next chapter. For now, we're going to focus on rocky reefs, which are not initially constructed by living organisms. However, rocky reefs often share space with tide pools and kelp forests, and are shaped by their inhabitants as much as by the underlying structure. Rocks provide substrate for the holdfasts of giant kelp, which need a sturdy base from which to reach toward the sun. Smaller encrusting algae grow over rocks, while many animals actually bore into rock, creating new living spaces for both themselves and other opportunistic species.

↑ Seagrasses (inset) provide food and habitat, oxygenate water and sediment, sequester carbon, and protect coastlines from flooding and erosion. Although the structure of rocky reefs such as this one off the coast of Madagascar (top) is created by geology, not biology, the rocks are filled and covered with a profusion of encrusting, boring, clinging, crawling, growing life.

Why plants (and algae) matter to predators

Cephalopods don't eat plants. So why would they hang around seagrass and kelp? It's because organisms that make their own energy, called autotrophs, are the foundation of the food chain. (In the deep sea, we'll meet autotrophs that are not photosynthetic, but up here near the surface they're all sun-powered.) Not only do these autotrophs build an environment in which cephalopods can hide and crawl and swim, but they feed the grazers, which then feed the cephalopods.

MEET THE KELP

Giant kelp (*Macrocystis pyrifera*) and bull kelp (*Nereocystis luetkeana*), which create forests along the Pacific coast of North America, both grow at an astonishing speed—up to 1½ ft (0.5 m) a day. They can accomplish this thanks to abundant nutrients in the water, which are brought to the surface by upwelling. Upwelling occurs in temperate parts of the ocean, when currents push warm surface water offshore and draw cooler water that is richer in nutrients up from the depths.

Land plants take up nutrients through their roots, but kelps can take up nutrients over their whole surface area. Their holdfast, which resembles a root ball, exists solely to live up to its name—holding fast against currents, waves, and storms. From the holdfast grows a long stipe (a stalk that resembles a stem or trunk), and on this stipe is either a single large gas bladder (*Nereocystis*) or many small gas bladders

← The gas bladders that lift kelp fronds toward the sun are filled with gases produced by the surrounding cells.

→ Brown algae are harvested for the commercial production of alginate, a substance that can absorb, gel, and thicken.

(*Macrocystis*) that serve as floats to hold the kelp up in the sunlit region. From the gas bladders grow fronds, or blades, which are the main surface for photosynthesis.

SEAGRASSES, IMPORTANT AND VULNERABLE

Seagrasses could seem like a minor player in the ocean, as there are only 72 species of seagrasses compared to over 10,000 species of land grasses. But these seagrasses grow in marine meadows expansive enough to support huge grazing mammals (sea cows were aptly named) and dense enough to form a protective habitat for the vulnerable young of many species.

Seagrasses are less abundant, diverse, and resilient than algae. Seagrasses can't survive big waves; giant kelps are routinely torn up by winter storms, then rebuild their forests. Algae, especially the single-celled kinds, can live at extremes of temperature and desiccation. Single-celled algae called diatoms are so abundant that their skeletons, raining to the seafloor over millions of years, created the massive deposits of chalk that were later raised up by plate tectonics as the white cliffs of Dover on the southeast coast of England. Seagrasses, by contrast, tend to be picky life-forms, although they do have the ability to grow in brackish water that is a mix of fresh and salt water.

KELP CAMOUFLAGE

Many animals live directly on the kelp itself. The California Lilliput Octopus (*Octopus micropyrsus*) is described primarily from *Macrocystis* holdfasts. These octopuses grow to an adult mantle length of 1¼ in (3 cm), so a large holdfast of 3¼ ft (1 m) provides a perfect microhabitat for hiding, hunting, and laying eggs. Within, around, and beyond kelp forests are many smaller algae that also offer valuable habitat and camouflage opportunities—this includes the red algae mimicked so effectively by the Algae Octopus (*Abdopus aculeatus*, pages 62–63).

Estuaries and fresh water

In addition to the interface of land and sea at the intertidal, the edges of the sea create another, more subtle, interface between two habitats: fresh water and salt water. The sea is a receptacle for all the rivers and streams around the world, so why, one might wonder, is it salty at all? Every source of water that fills it is fresh, including rain.

→ Utah's Great Salt Lake has been slowly drying up. The reduced volume of water becomes saltier, posing risks to the survival of algae, shrimp, and other organisms that live there—as well as the birds that depend on them for food.

↓ Myanmar's Irrawady Delta occurs where several rivers meet the sea. As their water channels widen, they move more slowly, dropping sediments and creating a rich plain for rice cultivation.

PUTTING THE SALT IN THE SEA

However, when we examine all these water sources more closely, we see that they do contain small traces of salt, along with other minerals. The salt we usually think of, both in the sea and on the table, is sodium chloride. A chunk of pure sodium chloride is called "halite," by the same naming convention that makes a big chunk of calcium carbonate "calcite" or a big chunk of iron sulfide "pyrite." However, most rocks on Earth are not chunks of single minerals, but combinations of many. As rain falls on rocks and rivers flow over rocks, they weather these materials, breaking off little bits through means both mechanical (physical agitation) and chemical (acidic dissolution).

The water remains "fresh" because the amount of minerals dissolved in it is so small. Billions of years ago, when the future ocean basins of Earth first began to fill, their water would also have been fresh with only traces of minerals. Over millions and millions of years, the oceans became salty due to evaporation. The surface of the ocean is vast, exposed to the wind and heated by the sun, encouraging a phase change in water from liquid to gas. When this happens, the water molecules release their hold on one another and bounce around freely, floating as gas and eventually condensing as clouds, high in the atmosphere. When the water evaporates, dissolved minerals stay behind. This cycle has continued for much of the planet's lifetime: the input to the ocean from land contains a small amount of minerals, but the ocean's only output contains none, and so the sea has steadily built up its salinity.

Salt is not the only mineral in the ocean, though it's by far the most prevalent one, and the one we taste. Calcium is another important mineral in the sea, used by corals and other shell-builders to make their skeletons. Is this why both corals and cephalopods cannot live in fresh water, because they require the calcium in the ocean? No, there is also calcium dissolved in fresh water—enough for freshwater snails and clams to build their shells. Corals and cephalopods and other exclusively marine animals simply can't cope with the absence of salt.

THE SEA IN OUR VEINS

Life first evolved in the sea, and although we don't know exactly when or where it happened, we're pretty sure that the water was already salty when it did. The first single-celled life-forms may even have evolved in an extra-salty environment, such as shallow pools at the edge of the sea where water evaporates quickly, or deep hydrothermal vents where extra minerals are pumped into the water from Earth's innards.

We all carry a legacy of salt water inside our cells. The fluids in our bodies are salty; not only our tears and blood but the liquid inside and between our cells. Animals that still live in salt water have it easy because the liquid outside their bodies and the liquid inside their bodies are roughly equivalent. If they want to change the concentration of minerals inside their body,

they have to expend a lot of energy—for example, a nautilus building its shell or a cuttlefish building its cuttlebone has to work hard to collect a high concentration of calcium in a small volume of water so it can precipitate calcium carbonate into shell material.

TOO MUCH OF A GOOD THING

Animals outside of salt water have a challenge. We've already talked about the difficulties on land of keeping yourself moist and not drying out. Animals in fresh water have the opposite situation—they need to prevent themselves from absorbing too much water! The problem is a process called osmosis. When chemicals are dissolved in water in different concentrations on opposite sides of a permeable membrane (like skin), they will move across until the concentrations on each side are roughly equal. Osmosis occurs when the chemicals themselves cannot move across the membrane, so the water moves across instead, diluting the side of higher concentration until the concentrations are equal. The process of fresh water diffusing into a salty cell can continue until the cell membrane bursts!

Living cells in fresh water need a mechanism to either prevent osmosis or pump the water out as fast as it diffuses in. Both options take a lot of energy, and the more permeable the animal, the more effort it takes. Freshwater fish have scales to limit diffusion across their skin. Freshwater mollusks such as snails and clams can hide from fresh water in their hard shells. But cephalopods have a large surface area of extremely permeable skin. They are also active swimmers, already expending large quantities of energy on moving and hunting. That is thought to be the explanation for why there are no freshwater cephalopods. It's simply never been advantageous enough for them to move into fresh water and adapt to handle it.

A few cephalopod species can live in brackish water, where fresh water mixes with salt water at the mouths of rivers. Researchers are exploring how these species cope with this reduced salinity, and it's possible that future research will reveal other limitations that have prevented cephalopods from truly colonizing fresh water.

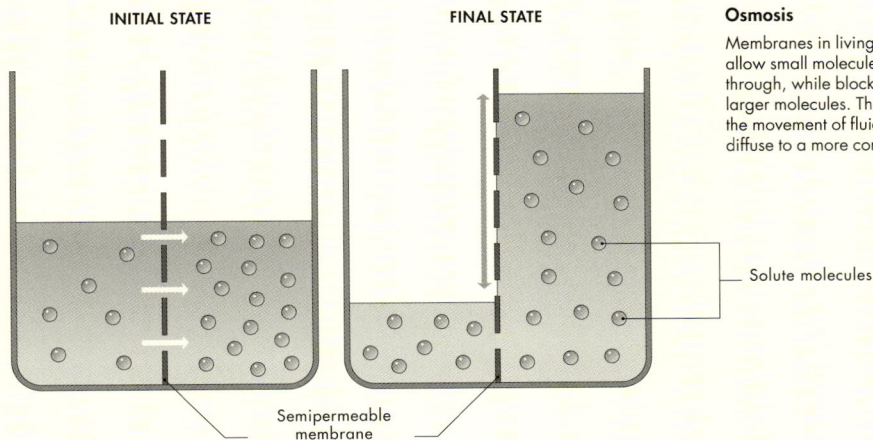

↗ Trout and other freshwater fish take up water through osmosis, but instead of expanding until they burst, they have adapted to divest themselves of this excess water by producing abundant dilute urine.

↖ Where air meets water, more oxygen is often available, but tissues can dry out. Where fresh meets salt water, more nutrients are often present, but osmotic challenges loom large.

INITIAL STATE

FINAL STATE

Osmosis

Membranes in living organisms typically allow small molecules, like water, to pass through, while blocking the movement of larger molecules. This leads to osmosis, the movement of fluid from a more diffuse to a more concentrated solution.

Solute molecules

Semipermeable membrane

Effects of nearshore pollution

Pollution of many kinds can accumulate in reef, kelp, and seagrass environments. Trash enters the sea when it's dropped from boats, washed off the shore, or blown out from land on the wind. These things can all happen accidentally and are rarely the result of malicious intent, but the results are inescapable: garbage where it shouldn't be, tangled in kelp or seagrass, lodged in rocks or corals, even carried by currents to the most remote parts of the ocean.

RUNOFF AND ALGAL BLOOMS

Chemical discharge is also a marine issue. Chemicals include anything from heavy metal manufacturing effluent to agricultural fertilizer runoff. Fertilizers at first might sound like a beneficial input for kelp forests and seagrass beds. The problem is that they offer far more nutrients than the ecosystem evolved to handle. Rather than stimulating the growth of large algae and plants, the fertilizer is used by single-celled algae in the plankton. These algae bloom in a population explosion, then die and sink at the end of their brief lives. This bonanza of food doesn't support the existing grazers of kelp and seagrass, but is devoured by decomposing microbes that use up the oxygen in the water in their digestive process. (Oxygen is also used up by the algae themselves during the night, when they cannot photosynthesize.) Runoff leads to algal blooms, which lead to deoxygenation, which leads to die-offs of fish and other large animals due to asphyxiation.

Algal blooms are not always a result of human inputs throwing the system out of balance. Nutrient availability typically fluctuates in the sea, and blooms are a natural process that has simply become more frequent and pronounced as a result of industrialized agriculture.

OIL AND WATER

Another source of pollution that can wreak havoc on marine systems is oil. Oil spills tend to make dramatic headlines, accompanied by eye-catching photographs

of oil-soaked birds, mammals, and fish dying on the beaches, with heroic humans fighting to rescue them. This attention focuses on the acute problem, but there are lesser-known chronic complications that can last for many years after an oil spill. Dispersed and broken down into microscopic pieces, the oil remains in the system, disrupting animal development and affecting generation after generation.

Spills can happen anywhere that oil is drilled or transported, but nearshore areas are the most affected (the open ocean and deep sea have pollution issues too, see pages 158–159). Boat traffic is greatest along the coasts, and in addition to the risk of spilling oil and other dangerous chemicals, boats can physically damage the ecosystem, scraping organisms off rocks and tearing up kelp and seagrasses.

↗   An algal bloom viewed from above shows how microscopic organisms can multiply to cover expansive areas.

→   Tankers move 1–2 billion tons of crude oil across the sea every year. If they run aground or are damaged, the spills can be catastrophic.

←   Plastic in the ocean can damage and kill animals from the outside, if they become entangled, or from the inside, when they mistake it for food and consume it.

To find a den or build a den?

The classic octopus is a benthic octopus. This is a descriptor for organisms that live on or in association with the seafloor, whether it's sand, mud, rock, or so overgrown with algae you can't be sure what's underneath. Marine animals that aren't benthic can fall into one of two categories: neritic or pelagic. Both words describe animals that swim up in the water rather than settling on or crawling on the ground. Neritic animals swim near the shore, over the continental shelf, like the squids in this chapter's profiles. Pelagic animals, meanwhile, swim or drift in the open ocean, far from forests, meadows, and reefs. We'll encounter pelagic cephalopods in the open ocean (pages 144–177).

OCTOPUSES AND THEIR DENS

A propensity for making and keeping a home is a common feature of benthic octopuses, and one of the ways in which they are most relatable to us humans. There are many approaches to acquiring a den. Structures created by rocks and algae provide plenty of hiding holes. Empty seashells are suitable for small octopuses, and bottles and cans discarded by humans are perfectly serviceable as well. A 2022 study identified an increased incidence of octopuses using trash as dens, and noted that the recently described pygmy octopus of Brazil (*Paroctopus cthulu*) has never been seen using natural shelters but is mostly reported using empty beer cans.

Octopuses can be competitive when it comes to space, and as carnivores they may eat a snail or clam and then take up residence in its shell. Generally, though, octopuses tend to eat prey smaller than they are, so the shells thus emptied would not be large enough to live in. Instead, many octopuses pile the empty shells outside of their dens into a midden. (Shellfish-eating humans have made middens

throughout history, too.) With their strong and dexterous arms, octopuses can move shells and rocks and sand to excavate or construct a den. They may use a blast of water from their siphon to clear sediment from under a rock, creating space to squeeze their body inside. They may also steal dens from other octopuses.

DENNING TOGETHER

The Common Sydney Octopus (*Octopus tetricus*, pages 108–109) has garnered media attention for gathering in high-density housing (by octopus standards) off the coast of Australia. In 2009, divers reported a site dubbed "Octopolis," where an area full of shells had become a center of octopus life for multiple octopus generations. The middens had grown to cover the area in layers of shells, but underneath researchers were able to identify an object about 1 ft (30 cm) long that may have attracted enough octopuses to kick off the settlement. Over the years since, divers have found Octopolis continuously occupied, with a population ranging from 2 to 16 individuals.

In 2017, a second populous site of Common Sydney Octopuses was discovered and named "Octatlantis." No single seed object was found this time, but possibly some rocks had attracted a few octopuses, and as they denned and brought shells and built middens, the area grew to be a magnet in a relatively structure-free area. Octopuses in both areas interact aggressively over available dens, defending them from potential takers.

↖ Octopuses often use their water jets to push away undesired objects, both animate and inanimate.

↑ A well-camouflaged octopus nestles in a midden, or collection of empty shells. Often octopuses construct their dens partially or completely from the leftovers of their meals.

← This octopus hides its soft, vulnerable mantle inside its den while keeping an eye on its surroundings. Suckers along the edges of its foremost arms adhere to the rock.

↑ Octopuses like this Coconut Octopus that live near coastal human habitat often make use of litter for shelter. They risk injury from the sharp edges of broken objects, as well as well as possible health impacts from toxins leaching into the water.

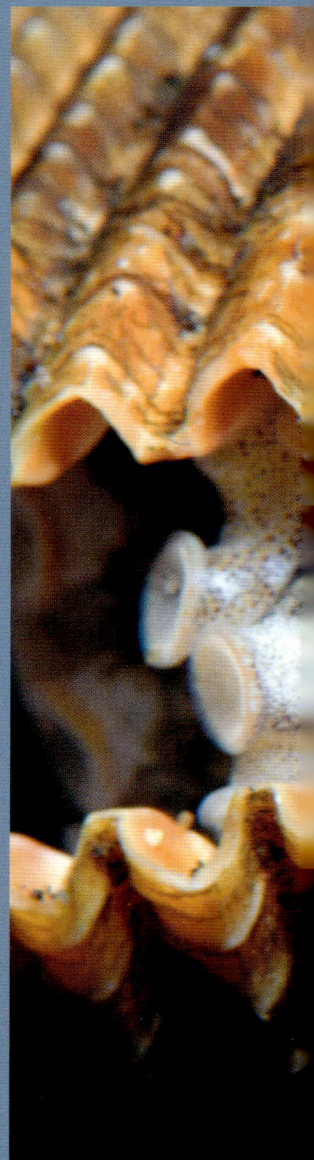

↗ Having long lost their ability to construct a shell, octopuses have adapted to make use of the shells of their mollusk cousins. This Coconut Octopus may have eaten the clam that made this shell or it may have encountered the shell while exploring.

Gathering to mate and spawn

Finding a mate is arguably the most important life task of any sexually reproducing organism. If evolutionary success is defined by passing your genes on to your offspring, and you can only produce offspring by mating, then making that connection is of paramount importance.

GIANT CUTTLEFISH SPAWNING

Several species of cephalopods gather in regular seasonal mating aggregations in nearshore locations. The Giant Cuttlefish (*Sepia apama*) in Australia is one such animal, and serves as a tourist attraction for the nearby town of Whyalla. The Giant Cuttlefish is also one of the most freshwater-tolerant cephalopods, meeting its mates in an estuarine bay. Its habitat is a combination of seagrass beds and rocky reef.

Giant Cuttlefish are the only cuttlefish known to aggregate seasonally for mating. By contrast, such gatherings are the most common reproductive strategy of nearshore squid (Loliginidae), such as the Slender Squid (*Doryteuthis plei*, pages 110–111) and the Luminous Bay Squid (*Uroteuthis noctiluca*, pages 112–113).

MATING STRATEGIES

When cephalopods come together to mate and spawn, their interactions are typically driven by competition between males for access to females. This type of mating system is well known in many other animals, and usually arises when females invest more energy than males in their offspring. While female squid and cuttlefish do not guard their eggs like female octopuses do, they must invest significant resources in producing yolk-rich eggs and the rest of the capsule material.

This competition has resulted in diverse male reproductive strategies. In some species, males can be

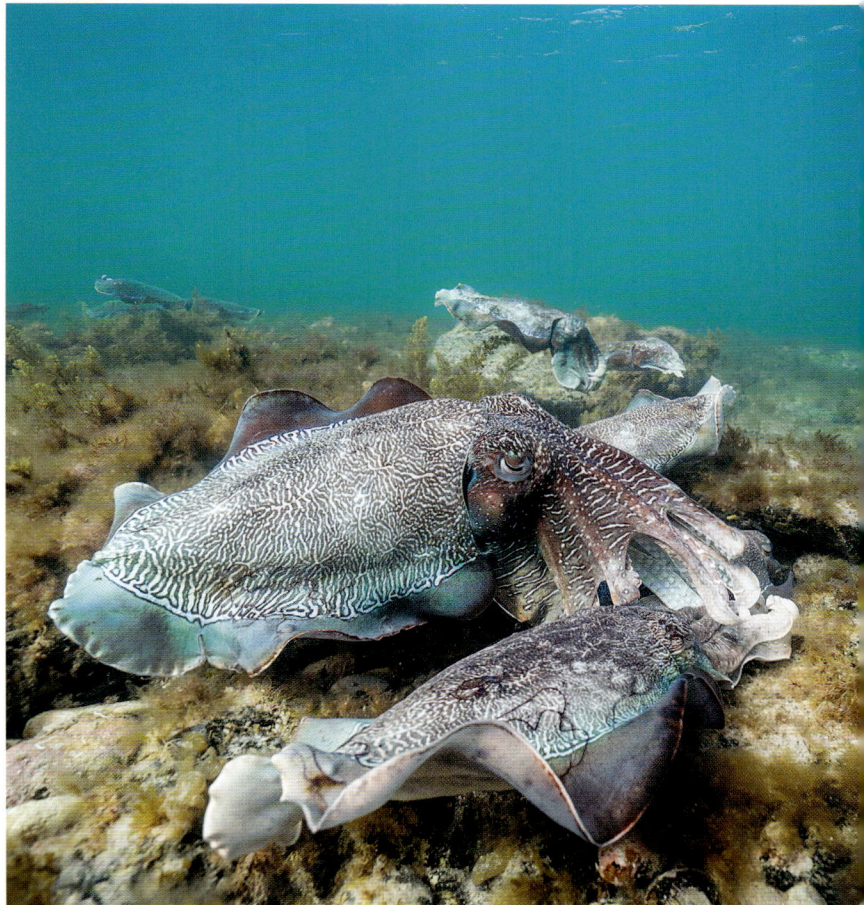

→ These Giant Cuttlefish living off the southern coast of Australia signal their readiness to mate with intricate color patterns on their skin.

← Opalescent Squid gather to mate off the coast of California. Male arms flush red as they grasp the mantles of females and deliver sperm. Females produce egg capsules, which they attach to the seafloor.

identified as either "consort males" or "sneaker males." Consort males are larger and more aggressive, mating with a female and guarding her to prevent her from mating with other males, while sneaker males are smaller and try to mate with females when the consort males aren't looking. Sneaker males will sometimes even use their camouflage abilities to disguise themselves as females, which allows them to slip past the notice of consort males.

Sneaker and consort males each produce different spermatophores (packages that contain sperm) *and* the sperm inside the spermatophores is different as well. These are not genetically determined differences—one male can change as he grows from producing sneaker spermatophore/sperm and engaging in sneaker behavior to producing consort spermatophore/sperm and engaging in consort behavior.

Far less is known about the flip side of reproductive behavior: female choice. Researchers are just beginning to examine how females choose which mates to accept, gather spermatophores from multiple males, store them, and finally select which to use for fertilization at the time of egg laying.

Gathering in a mating group is advantageous for finding a mate, but also provides an easy mass target for predators. It's a race to mate and spawn before you're captured by hungry fish or mammals—including humans.

Targeted by fishers

Nearshore cephalopods are commonly fished and eaten by people. The loliginid squid *Doryteuthis opalescens*, for example, is even called the Market Squid. Unlike the cephalopods featured in the last chapter, most of these species aren't so close to land that people can pick them up from tide pools, but they can be easily collected by divers or fishers in small boats with nets or poles. Commercial fisheries also target nearshore species with much larger boats.

FISHING TECHNIQUES

Since many of these species move toward the surface at night, cephalopod fishers often work during the dark hours. Artificial lights can entice squid in greater numbers closer to the surface, possibly by attracting prey species that the squid can then hunt. Many squid fishing fleets employ "light boats," and the glow of these boats can be seen in satellite photos of Earth.

One of the most common techniques for catching squid is called "jigging." This kind of fishing can be done with a pole or a hand line or even a machine, as long as the end of the line bears a jig. A squid jig is hook, weight, and lure all in one, and typically glows in the dark. It is shaped like a torpedo with concentric circles of hooks that entangle the suckers and appendages of a squid that strikes at the jig as if it were prey.

Cephalopod fisheries tend to be less regulated by governments than other fisheries, perhaps in part because we haven't yet managed to fish any cephalopod population to extinction. By contrast, we've learned the hard way that we can wipe out other marine life such as oyster beds and sea cows.

FISHERY MANAGEMENT

Many cephalopod fisheries do have quotas—limits on the quantity of animals that can be caught in a given year or season. Setting a quota presents a significant challenge for managers, however, because cephalopod population sizes are unpredictable compared to many other organisms. Their quick generation time and large number of eggs lead to frequent alternation between boom and bust, with the population expanding and

← This squid jig is shaped like a shrimp to entice squid that feed on such prey. Instead of hooking the squid through the mouth, as barbs designed for fish often do, a jig catches a squid by the suckered arms.

contracting due to a confluence of different factors, of which fishing pressure is only one.

An additional complication arises for those squid that are targeted by fishers over their spawning grounds. If every single squid or cuttlefish is able to reproduce before being caught, then fishing would have no direct impact on cephalopod population size. (It would still impact the populations of predators and scavengers that usually eat those animals during and after spawning, though.) On the other hand, if all of the squid or cuttlefish are caught before they can mate and lay eggs, then that's it—the population or species is gone. Obviously, the catch will always fall somewhere between these extremes, and it's difficult to calculate quotas and seasons to balance it out.

Overall, most cephalopod populations for which we have data are healthy, and many are even growing. This could be due to a release of predation pressure from other species that humans have overfished, or it could be a positive response to climate change. Or it could be something else entirely.

↑ This squid fishing "light boat" attracts squid to the surface, while a companion boat or boats will cast nets to collect the squid.

↗ Lost or abandoned fishing gear clutters the sea, sometimes trapping marine life. This net, however, has also served as an egg-laying substrate for squid.

OCTOPUS SP. UNKNOWN

Hairy Octopus

Frilled mystery

SCIENTIFIC NAME	*Octopus sp.*
FAMILY	Octopodidae
MANTLE LENGTH	¾ in (2 cm)
TOTAL LENGTH	4 in (10 cm)
NOTABLE ANATOMY	Long, tufted papillae
MEMORABLE BEHAVIOR	Algae mimicry

As was the case for the Atlantic Longarm Octopus (*Macrotritopus defilippi*, pages 68–69), it's likely that when the Hairy Octopus receives a scientific name, it will leave the genus *Octopus*. For now, however, the species has barely been studied scientifically; it is known almost exclusively from photographs taken by recreational divers.

In 2000, only a single individual had been documented. It appears in the book *Cephalopods: A World Guide* as "Octopus sp. 16" and the author, Mark Norman, reports that it could be a juvenile or an adult. Over the following two decades, underwater digital photography technology has improved, an expanding human population has led to more divers underwater, and communications technology has made it easier to exchange information about where to find attractive and unusual species. We now have many more photographs of Hairy Octopuses, but not a single scientific paper has yet been published on the species.

Although placed here in the rocky reef section, the Hairy Octopus has also been observed in coral reef habitats

(pages 114–143). Its camouflage is similar to, but even more dramatic than, the Algae Octopus (*Abdopus aculeatus*, pages 62–63). The Hairy Octopus makes use of papillae—folds of skin and connective tissue created by the activity of muscles squeezing to make skin stretch or bulge. It's a little bit like holding a water balloon in your hands—if you clench your fist around the middle, then water-filled bumps will push out on either side. However, the muscles in the octopus's skin have much finer-grain control over the shapes, sizes, and locations of the bumps.

It is the texture of their skin that is most striking, but Hairy Octopuses still have and use their chromatophores, varying in color from red to brown to white, and in pattern from solid to spotted. They seem inclined to change the color of the skin around their eyes most frequently, switching from red to white and back again in flashes, waves, or ripples. This could be an example of a disruptive eye mask, a type of camouflage that's reasonably common among both predators and prey. Eyes are often the most easily spotted part of an animal, so making them difficult to identify is a kind of camouflage shorthand.

→   This Hairy Octopus was photographed off Ambon Island, part of a volcanic archiphelago in Indonesia.

OCTOPUS CHIERCHIAE

Lesser Pacific Striped Octopus

Multiple mother

SCIENTIFIC NAME	*Octopus chierchiae*
FAMILY	Octopodidae
MANTLE LENGTH	1 in (2.5 cm)
TOTAL LENGTH	1½ in (4 cm)
NOTABLE ANATOMY	High-contrast stripes
MEMORABLE BEHAVIOR	Not dying after laying eggs

This small resident of the eastern tropical Pacific (off the coast of Central America) has rocketed to fame in the cephalopod science community because of its life cycle, which is unusual for an octopus. Female Lesser Pacific Striped Octopuses continue to feed and grow while caring for their eggs, and instead of dying when their babies hatch, they go on to repeat the process multiple times.

Males have some distinctive behavior to show off as well— when courting females, they shake and twirl the tips of their arms. Although the exact purpose of the behavior is unclear (is it attracting attention, communicating interest, displaying physical vigor, or a combination of all these?), it's clearly a type of courtship behavior, something that's also rather unusual among octopuses, but more common among squid and cuttlefish.

The ability of Lesser Pacific Striped Octopuses to mate and lay eggs multiple times makes them ideal laboratory organisms, since many experiments require raising multiple generations of octopuses. Furthermore, they offer a contrast to other octopus mothers' early deaths, opening opportunities for comparative studies of life cycles and senescence (age-associated deterioration).

The fact that this species can reproduce more than once— a feature called iteroparity, rather than exhibiting the semelparity, or reproduce–then–die lifestyle of other octopuses—has been known since 1984. However, the animals were difficult to find in the wild and few laboratories had the resources to try to rear them. Now, at a state-of-the-art laboratory in Woods Hole, Massachusetts, where a team of scientists is dedicated to keeping cephalopods alive and healthy, the Lesser Pacific Striped Octopus has become a promising model system.

This species' common name alludes to the existence of the Larger Pacific Striped Octopus. The two species overlap in both distribution and habits and, unfortunately, in initials. While both could be referred to as LPSOs, this initialism is used exclusively for the larger of the two species. This might be because the larger species has proven more difficult to study, and still lacks a scientific name. The LPSO, like its smaller relative, exhibits unusual reproductive habits—a risky beak-to-beak mating position, and mating pairs sharing dens.

→ Although Lesser Pacific Striped Octopuses can now be raised in captivity, their particular tastes challenged scientists trying to get to that point. Among other needs, these animals apparently prefer grass shrimp to brine shrimp.

Southern Pygmy Squid

Wee glue-maker

SCIENTIFIC NAME	*Xipholeptos notoides*
FAMILY	Idiosepiidae
MANTLE LENGTH	1 in (2.5 cm)
TOTAL LENGTH	1¼ in (3 cm)
NOTABLE ANATOMY	Glue gland
MEMORABLE BEHAVIOR	Sticking to seagrass

Pygmy squids (Idiosepiidae) are the smallest adult cephalopods in the world, with males maturing at significantly smaller sizes (a maximum of ⅝ in/1.5 cm) than females. They've evolved a unique lifestyle along with their tiny size.

Rather than chasing their prey, pygmy squids are sit-and-wait predators. They have a glue gland for attaching to seagrass leaves and, unlike small cuttlefish that use suction and glue to prevent themselves from being swept away, the primary aim of pygmy squid glue is to keep the animals hidden from view. This may be why they evolved their glue gland on the upper surface of their mantle rather than the bottom. This allows them to stick to the underside of seagrass leaves, where they are nearly undetectable.

When a shrimp draws within range, pygmy squids can instantly release their glue, probably by producing an unsticking chemical. They can then jet toward the shrimp, grab it, and strike with their tiny, but still deadly, beak. Like many other cephalopods, they aim directly for the central nervous system with the first bite, to render the prey immobile as it is consumed.

Once placed in the genus *Idiosepius*, the Southern Pygmy Squid was moved to its own genus, *Xipholeptos*, in 2018, based on genetic and anatomical divergence from other members of the *Idiosepius* genus. It lives in the Pacific Ocean off southern Australia, where seagrasses are both abundant and threatened. In and among the fronds lurks an incredible diversity of tiny organisms, from those that remain tiny all their lives, such as pygmy squid and the shrimp they eat, to those that are only tiny at the beginning of their lives, such as young fish that will eventually grow to be open ocean swimmers.

The Southern Pygmy Squid is large for a pygmy squid, compared to its tropical cousin the Two-toned Pygmy Squid (*Idiosepius pygmaeus*). This is consistent with the trend of animals in cooler water getting bigger and living longer. (The Pacific off southern Australia is more temperate than the warm Indo-Pacific habitat of *I. pygmaeus*.) Females grow larger than males in both species; in the Northern Pygmy Squid (*I. paradoxus*) it's because they grow faster, but in the Southern Pygmy it's because they live longer. Females can live more than 100 days, males half of that, which leads to a situation known as "cross-generational spawning."

Tiny squid

Adult pygmy squid are some of the smallest cephalopods in the world.

Male at actual size (15 mm)

→ Pygmy squid have a distinctive use for their ink. When approaching prey, they will first squirt an ink smokescreen to hide their movement, then attack from behind it.

Pacific Red Octopus

Ruby rock-lover

SCIENTIFIC NAME	*Octopus rubescens*
FAMILY	Octopodidae
MANTLE LENGTH	4 in (10 cm)
TOTAL LENGTH	15¾ in (40 cm)
NOTABLE ANATOMY	"Eyelash" papillae
MEMORABLE BEHAVIOR	Abundance and flexibility as a predator

Relatively abundant for an octopus, this is the species most likely to be spotted while snorkeling, diving, or otherwise experiencing the eastern rim of the Pacific Ocean in the northern hemisphere, from Mexico to Alaska. The Pacific Red Octopus is sometimes called the East Pacific Red Octopus, since it's not clear whether this particular species is ever found in the western side of that ocean. Red is a fairly common resting color for octopuses of many species, but it is particularly noticeable for this species.

Pretty much anything that you might think of as a typical octopus trait applies to the Pacific Red Octopus. They are speedy changers of skin color and texture, and voracious predators of all kinds of shellfish from scallops to crabs. They reproduce once at the end of their lives, and their many tiny eggs hatch into planktonic paralarvae that drift with the currents before growing large enough to settle back down to the rocks.

Often confused with juveniles of the much larger Giant Pacific Octopus (*Enteroctopus dofleini*), the Pacific Red Octopus can be identified by one intriguing consistency in its skin: three little papillae underneath each eye, sometimes called eyelashes.

The species is relatively easy to keep in captivity, because it can be fed a variety of foods and requires only enough space for a den and a bit of roaming. Rocks and shells are its den material of choice, and it often builds up a midden outside its den. Like all octopuses, Pacific Reds can deliver a venomous bite—their venom is not strong enough to kill a human, but it is painful and takes a while to heal.

Because they live in cracks and crevices, they are sometimes collected by accident, as aquarists at the Monterey Bay Aquarium discovered when a Pacific Red Octopus traveled into an exhibit as a small juvenile hidden in the folds of a sponge. It made a significant dent in the crab population of the aquarium before it was discovered—walking across the floor! They are not as keen to crawl out of the water as the "land octopus" (pages 62–63), but like many subtidal octopuses, they will turn themselves into intertidal octopuses if the opportunity presents itself.

→ Pacific Red Octopuses are one of the species most likely to be casually encountered by humans, and are also inclined to bite when threatened (for example, when an enormous hand picks them up). Their venom is not fatal, but it is painful and bites can take weeks to heal.

MACROCTOPUS MAORUM

Maori Octopus

Big kiwi

SCIENTIFIC NAME	*Macroctopus maorum*
FAMILY	Octopodidae
MANTLE LENGTH	12 in (30 cm)
TOTAL LENGTH	3¼ ft (1 m)
NOTABLE ANATOMY	Large size
MEMORABLE BEHAVIOR	Lobster stealing

Found in the Southern Ocean off the coasts of New Zealand and southern Australia, the Maori Octopus is named for the indigenous people of New Zealand. It is one of the largest octopus species in the world, after the Giant Pacific Octopus (*Enteroctopus dofleini*) and the Southern Giant Octopus (*E. magnificus*).

This species' diet overlaps with that of nearby humans, as both are fond of rock lobster. The local lobster fishery is conducted by lowering baited traps into which lobsters climb and get stuck. This creates accidental bait for octopuses, which can climb both in and out of the traps, feasting on lobsters that cannot run away. While other predators also occasionally eat lobsters out of traps before they can be collected by fishers, lobsters lost to octopuses are recognizable because, as scientists report, they are always "carefully dismembered" rather than crushed or broken.

In addition to catching them accidentally in lobster traps, humans also fish for Maori Octopuses on purpose, as do plenty of other marine predators—at some risk to themselves. In 2017, a bottlenose dolphin was found dead on a beach with a Maori Octopus in its mouth, and a post-mortem examination confirmed that the dolphin had choked to death on its last meal. (The octopus didn't survive the encounter either.)

Seals and albatrosses are also eager eaters of Maori Octopuses, and their stomachs offer a perhaps surprising opportunity to figure out how big these animals can really get. Octopuses are particularly appealing prey items because they are nearly all digestible, with no shells or bones to contend with, but they do have one hard part: a beak. The indigestible beaks of cephalopod prey often accumulate in predator stomachs, and because there's a relationship between beak size and octopus size, these beaks can be used to estimate the size of the octopus prey that was consumed.

Of course, the relationship isn't exact. Taller humans tend to have longer feet, for example, but that doesn't mean a given foot size indicates an exact height. Still, there is an overall correspondence, and calculating octopus size based on beak size has confirmed that Maori Octopuses do not reach the same maximum sizes as the two giant *Enteroctopus* species.

→ These animals are called *wheke* in the Maori language. A mythical octopus named Te Wheke-A-Muturangi is said to have reshaped the land during a battle with a local warrior, creating the Marlborough Sounds on New Zealand's south island.

Common Sydney Octopus

Urban octo

SCIENTIFIC NAME	*Octopus tetricus*
FAMILY	Octopodidae
MANTLE LENGTH	5½ in (14 cm)
TOTAL LENGTH	2 ft (60 cm)
NOTABLE ANATOMY	Typically grayish coloration, dramatic white eyes
MEMORABLE BEHAVIOR	Communal living, throwing ability

The only octopus species known to live in multi-generational congregations sometimes referred to as "octopus cities," the Common Sydney Octopus is a poster child for studying social behavior in a group of animals generally considered to be solitary loners. At Octopolis and Octatlantis (see page 91), individual octopuses go about daily lives familiar to any urban inhabitant: hunting for food, avoiding predators, and squabbling with neighbors over space. Since the structure of the cities consists primarily of layers of shells the octopuses brought to the sites themselves, they are considered a kind of ecosystem engineer in this situation.

This species garnered its common name for its abundance in proximity to Sydney, Australia, but is also sometimes called the Gloomy Octopus for its often drab coloration. Its behavior, however, is anything but dull. At Octatlantis, studies revealed octopuses evicting each other from dens and trying to keep other octopuses out of the city entirely. At Octopolis, researchers recently recorded numerous cases of octopuses throwing dirt, shells, or algae at each other. (If you're an octopus, "throwing" means aiming the projectile with your arms, then impelling it with a jet of water from your siphon.) The unfortunate octopus targets sometimes ducked or raised their arms defensively. The throwers were often females rejecting mating attempts from males, but not always. Sometimes the throwers didn't target another octopus at all, but seemed to act out of displaced aggression—like an angry kid throwing a toy at the wall. And sometimes they threw objects at the tripod camera that scientists were using to gather data. The researchers described all this throwing as a form of tool use, as well as social interaction.

Common Sydney Octopuses may put on bland colors to match their fairly colorless environment of rocks, sand, and shells, but they're as capable as the next octopus of dramatic color changes. Researchers named one common intimidating posture the "Nosferatu display": when an octopus stands up on all of its arms, stretching tall, and quickly turns dark and ominous.

These "octopus cities" that have persisted over so many years of observation offer another opportunity to researchers: a chance to observe distinct personalities within a group of octopuses, finding that different individuals may have their own distinct den-building habits, for example.

→ This Common Sydney Octopus is surrounded by other species that have adapted to the same habitat in different ways: a bright orange sponge, a translucent sea squirt, and living coral.

DORYTEUTHIS PLEI

Slender Squid

Nearshore swimmer

SCIENTIFIC NAME	*Doryteuthis plei*
FAMILY	Loliginidae
MANTLE LENGTH	13 in (33 cm)
TOTAL LENGTH	19 in (48 cm)
NOTABLE ANATOMY	Long fins and short arms
MEMORABLE BEHAVIOR	Congregating and competing for mates

Sometimes also called the Arrow Squid for the shape of its body, the Slender Squid is a quintessential inshore species. These streamlined, active swimmers move in small groups, aggregate in larger groups to mate, and die after laying eggs. They live in the tropical western Atlantic from Brazil to Florida, and in at least part of this range they are considered a keystone species. This ecological term refers to a species that exerts a significant influence over the distributions and populations of many other species in the area.

Recent genetic research indicates that there may actually be two distinct species within *Doryteuthis plei*; one in the North Atlantic and Caribbean, and the other in the South Atlantic. The scientists who discovered this also figured out how and when the species might have split apart. About 16 million years ago, global sea levels were particularly low due to the amount of water locked up in large ice shelves. Between the North and South Atlantic, the entire continental shelf would have been exposed and dry, leaving no place for a nearshore squid to live. This would have isolated a previously contiguous species into two separate breeding populations, and over time they diverged enough that when they could again overlap, they were no longer interbreeding with each other.

In contrast to octopuses, where females are often larger than males, Slender Squid and most other inshore squid display the opposite sexual dimorphism, in which males are larger than females. For purely reproductive purposes, evolution usually favors larger females to make more and better eggs, since eggs take more energy and investment than sperm. However, in species where males compete for access to females, large male size often evolves as a way of winning the competition and securing mating rights.

Slender Squid have a life span of approximately one year, and they congregate in the summer to mate and spawn. They support a large commercial fishery, with the fleet catching squid both for human food and as bait to fish for other animals. Mating aggregations are the primary fishing target, and this creates a need for careful management in order to leave enough squid to successfully reproduce and maintain the species.

→ Red chromatophores are expanded on the skin of this Slender Squid, while yellows and browns are contracted and invisible. The animal's hydrodynamic shape has evolved due to the advantages of speedy swimming.

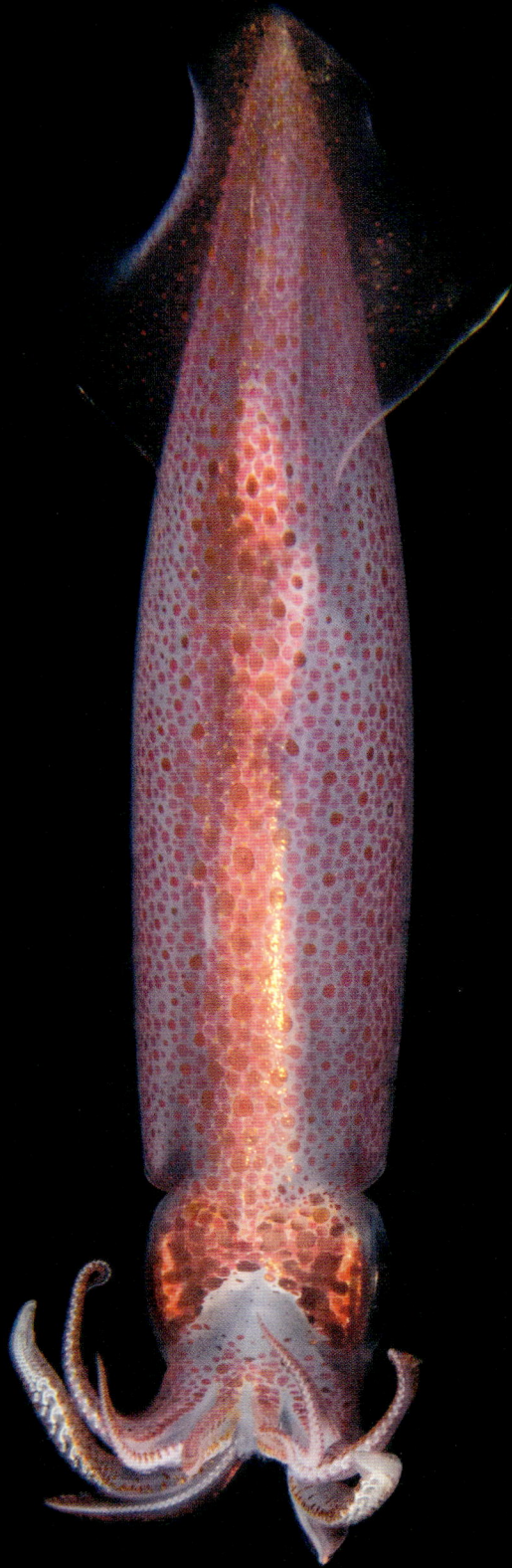

Luminous Bay Squid

Night light

SCIENTIFIC NAME	*Uroteuthis noctiluca*
FAMILY	Loliginidae
MANTLE LENGTH	3½ in (9 cm)
TOTAL LENGTH	5 in (13 cm)
NOTABLE ANATOMY	Bioluminescent organs
MEMORABLE BEHAVIOR	Tolerance of brackish water

This small eastern Australian squid is one of the few cephalopod species that can handle water less salty than the open ocean. It is sometimes found in estuaries, regions where rivers flow into the sea, mixing fresh and salt water to a degree that varies depending on the tidal flux. These are common places for seagrass beds, and indeed these squid seem to associate with seagrass—not for the grass itself, but for what they can find in it to snack on.

Because its range covers both tropical and temperate waters, the Luminous Bay Squid is a species that showcases the plasticity of the cephalopod life cycle. Where the water is warm, they grow fast and mature early, living for only four months. Females of this species are the larger sex, suggesting that although they belong to the same family as *Doryteuthis*, they engage in different reproductive habits—perhaps without male–male competition. They have not been studied enough to validate this.

As relatively small loliginids, Luminous Bay Squid are not targeted as a deliberate fishery, but show up as bycatch in nets aiming for prawns. The individual Luminous Bay Squid that find their way into a prawn catch must be temporarily in heaven, surrounded by a seafood buffet, but their delicate bodies are likely to be damaged by the experience even if they are thrown back in the sea alive.

Luminous Bay Squid have two light organs, which are similar in many ways to the light organs of bobtail squids in the genus *Euprymna*. Both make light on the underside of the squid's body for the purpose of counter-illumination—exactly underneath the dark ink sac, to block its shadow. Both also use bacteria from their environment to produce this light, all belonging to the *Vibrio* group. There are lots of *Vibrio* bacteria, including the human cholera–causing *Vibrio cholerae* and the fish pathogen *V. harveyi*. Luminous Bay Squid have evolved to use *V. harveyi* itself, along with some of its relatives, in their light organs, with no known ill effects.

Noctiluca means night light, which is lovely enough, but before being reclassified as *Uroteuthis*, this species also had a wonderful genus name: *Photololigo*, or "light loligo."

→   The genus *Uroteuthis* is the only loliginid genus with a bacterial light organ (it contains 14 species, all of which glow). One small group of squid appears to have evolved from this genus and lost their light organs in the process, for reasons that have yet to be determined.

CORAL REEFS

Summer in the city: coral reefs, tropical water, and global warming

The word "reef" simply describes a solid underwater structure. A reef can be made of rocks or corals or even sunken ships. Reefs of all kinds are significant to aquatic life, because they change a sandy or muddy seafloor into a three-dimensional living space, opening up new niches and increasing the surface area for all kinds of organisms to settle and grow.

CORAL ANATOMY

Coral reefs offer the added complexity of being made of living organisms themselves. Corals are animals, in the same group as sea anemones and jellyfish (phylum Cnidaria). Cnidarians have two primary body types: the medusa (which looks like a typical jellyfish) and the polyp (which looks like a typical anemone). Many, but not all, cnidarians alternate between the two forms during their life cycle. Looking closely at a coral, you can see that it is a colony of polyps. (There are also solitary corals, which are singular polyps, but these do not form reefs.) Each polyp is a tube with a ring of tentacles around its mouth. The tentacles can capture food, which goes into the mouth and down to be digested in the stomach. Any waste is then expelled through the same opening, which now serves as an anus. This system is called a "two-way gut."

A coral reef is a structure that can both eat and be eaten, kick-starting the local food web in a way that a rocky reef does not. Also, unlike rocks, corals grow. Each polyp creates an external skeleton, using materials from the surrounding water in a similar way to a mollusk building its shell. The coral takes a combination of calcium and carbonate that are dissolved in the water

← A crown-of-thorns starfish moves across coral colonies, extruding its stomach to digest polyps externally, then withdrawing both stomach and nutrients back into its body.

↙ Some coral colors come from algae that live inside the animals, but the polyps themselves can also produce suncreen pigments that block UV light.

↓ These coral colonies are releasing their gametes (eggs and sperm) in a coordinated spawning event that scientists are still working to understand.

and brings them together in high concentrations, producing a layer of calcium carbonate to add to its skeleton.

CORAL REPRODUCTION

As the colony grows, each new polyp adds its skeleton to the larger structure. But where do new polyps come from? Colonial corals grow by asexual reproduction. Each polyp can either bud off small polyps, or divide in half like an amoeba. Thus, a colony of coral polyps is actually a colony of clones. Since the polyps share the same genes and the same skeleton, this raises the philosophical question of what exactly constitutes an individual—is it the polyp, or the colony?

Corals can also reproduce sexually. Some species brood their young, while many others engage in broadcast spawning, releasing unfertilized eggs and sperm into the water at coordinated times to maximize the likelihood of fertilization. When this happens, the water over a reef can be thick with coral reproductive cells, and plenty of predators appreciate the snacks. Fertilized eggs that escape predation develop into tiny planktonic larvae, which eventually settle down and become the founding polyps of new colonies.

← A coral polyp's tentacles can be filled with photosynthesizing algae that actively make food from sunlight, even as those same tentacles comb the water for morsels to bring back to the polyp's mouth.

→ Excess heat is the most well-known cause of coral bleaching, but other stressors, including pollution and cold, can also force polyps to push out their partners.

ALGAL PARTNERS

Corals provide food for others in the form of their own growing colonies as well as their eggs and larvae. But where do corals get their nutrition? We heard about their tentacles catching bits of food, but most shallow corals get the majority of their energy from sunlight—like plants and algae do.

In fact, they require a partnership with algae to do it. Symbiotic algae called zooxanthellae live inside coral bodies, sharing their nutritious photosynthesis products in exchange for a sunny place to live. Together with each other and with an undoubted plethora of bacteria, viruses, and other microbes that we still don't know about, a coral and its zooxanthellae are considered a holobiont—a functional organism made up of other cooperating organisms. This is an endosymbiosis like that which led to the evolution of algae and plants, but not (yet) so entangled. Zooxanthellae are not dependent on their coral hosts and can also live freely in the environment, and, although some corals inherit their parents' zooxanthellae, many others must acquire zooxanthellae from what's available in the surrounding water.

CORALS AND CEPHALOPODS

No cephalopods are known to consume coral polyps directly, but many species use corals both living and dead as dens, places to hide from predators, and to lay their eggs. Because so many other organisms also use these spaces, competition and risk can be high. This may have led to the evolution of an unusual behavior of male Bigfin Reef Squid (*Sepioteuthis lessoniana*, pages 138–139) in which they probe crevices in advance of their mated females laying eggs in these locations. Scientists who observed this behavior speculated that the males were ensuring the safety of the egg–laying spot, a paternal investment in the care of their young that is unheard of in other cephalopods.

Many prey items of cephalopods are coral predators (sometimes called "coral grazers" because they defy the stereotypical image of a predator chasing down mobile prey and instead simply nibble on polyps that are stuck in position). Cephalopods in coral reefs eat many crabs and snails that eat corals, helping to contribute to a balanced ecosystem.

CORAL BLEACHING

The many brilliant colors of corals are part of what attracts humans to them, and we feel an aesthetic as well as an intellectual loss when corals "bleach," or expel their colorful cohabitors. Bleaching is a response to heat stress. It may seem counterintuitive to get rid of your main food supply when you're stressed out, but it is a short-term safeguard to the health of the coral. High temperatures cause zooxanthellae to produce "reactive oxygen species." These chemicals are the reason why oxygen was such a toxic compound back when cyanobacteria started producing it, before life evolved coping mechanisms. (Humans protect themselves against cell damage from the same chemicals using antioxidants.) Corals get rid of the zooxanthellae to protect themselves. It's a quick fix that leaves them vulnerable to starvation. Species that can gather food with their tentacles are somewhat better off, and it is possible for corals to acquire new zooxanthellae from the water when and if conditions improve. Still, coral bleaching often leads to coral death.

As ecosystems, dead coral reefs can recover with the settlement of new larvae establishing new colonies, and the rest of the reef inhabitants following. But they can also be lost forever, especially if different kinds of algae grow over the coral skeletons. Some reef animals, including cephalopods, may be able to adapt to an algae-dominated system, but many others will not.

An evolutionary hotbed

Warm tropical water is fairly low in nutrients, compared to cooler locations with temperate upwelling. That's why large kelps don't grow in the tropics, and it's also why tropical water tends to be so clear, without the abundance of single-celled algae that can turn temperate water green and murky. Nevertheless, tropical areas host a tremendous variety of species. The clarity of the water itself facilitates the evolutionary diversification of visual predators such as cephalopods.

← Reef-building corals and humans have a lot in common—we both like warm water and sunlight. This overlap in habitat can sometimes put us at odds.

↙ The clear tropics can offer 100 ft or more of underwater visibility. Fewer nutrients in these seas limit the growth of the algae that often cloud cooler water.

↓ Many of the flashiest fish we see in aquariums (as in this shot from Ripley's Aquarium in Toronto) are reef-dwellers from the Indo-Pacific.

INTRODUCING THE INDO-PACIFIC

In particular, the Indo-Pacific, a region that stretches from Australia to China and contains many islands in between, is a global hotspot of biodiversity both on land and sea. It was on these islands that the naturalist and collector Alfred Russel Wallace (1823–1913) came up with the idea of evolution by natural selection, independently of Charles Darwin (1809–1882). The physical structure of the Indo-Pacific is conducive to evolutionary adaptation, with a profusion of islands separated by manageable distances. Obviously, islands matter to terrestrial species, but they matter to marine species, too. A huge number of organisms are adapted to life on the continental margins, shelfs, and slopes, and cannot survive in ocean basins. Each island that breaks up the expanse of a basin provides a ring of shallow habitat around itself, as well as a stepping-stone between other such habitats.

CEPHALOPODS AND REEF LIFE

In the Indo-Pacific, populations can easily be isolated long enough to evolve into new species, yet not so isolated that individuals can never disperse to a new location and establish a new population. Cephalopods have several such methods of dispersal at their disposal. As adults they can move by swimming or crawling, and as planktonic paralarvae they can be carried by currents. They can even engage in rafting, or moving from place to place when a piece of habitat is broken loose and transported, often by a disruptive force such as a storm. For example, if a piece of coral in which an octopus or several were living broke off a reef and was carried to a new location, then the coral and its passengers might establish new populations—and their descendants might diverge from their parent populations to the point of becoming new species.

Cephalopods are certainly among the groups that have proliferated into diverse species in the Indo-Pacific. Of the seven species of nautiluses, all are associated with coral reef habitat in this region. We don't have similar estimates for octopuses, squid, and cuttlefish; however, we can guess that it might be similar to the situation for fish. The Coral Triangle alone, which is the center of biodiversity in the Indo-Pacific, is home to the most species of reef fish in the world.

There may be some bias at play in our perception of diversity. We humans are also visual hunters, and it's easier (besides more comfortable) for us to identify and name species in shallow clear water than in deep murky water, so there may well be many more undiscovered species in temperate regions. Keeping that in mind, it is interesting to note that some of the most dramatic new cephalopod species discovered and described in

the twenty-first century came from the tropical Indo-Pacific, including the Wunderpus (*Wunderpus photogenicus*, pages 70–71) and the Mimic Octopus (*Thaumoctopus mimicus*), and a 2022 study identified the Indo-Pacific as a hotspot of cephalopod biodiversity.

← Wunderpus paralarvae are transparent planktonic drifters, illustrating how a single animal's life cycle can connect distinct habitats, such as open ocean and coral reef.

↖ Chambered Nautiluses can often be found near reefs at night, although during the day they occupy much deeper habitat.

↑ This Mimic Octopus is one of the most eye-catching reef cephalopods. Divers travel from around the world to observe and photograph it.

The Coral Triangle

The biodiversity of the islands in and around the Coral Triangle is as remarkable on land as in the water. Naturalist Alfred Russell Wallace developed the theory of natural selection (independently of Darwin) during his exploration of this area.

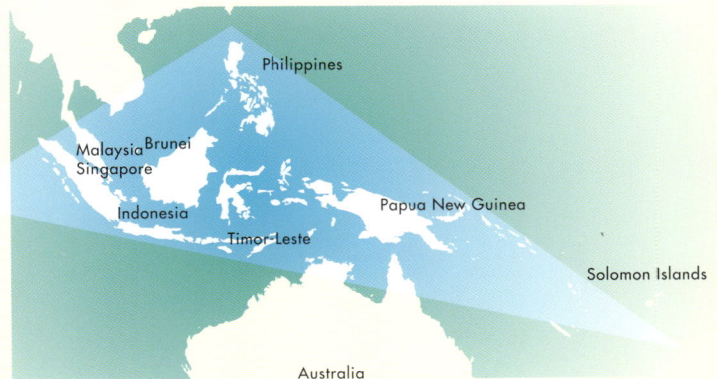

Philippines

Malaysia Brunei
Singapore

Indonesia

Timor-Leste

Papua New Guinea

Solomon Islands

Australia

Acidic seas and cephalopods

On land, climate change is often seen as synonymous with global warming. We know that certain places at certain times may instead grow colder, and we know that the effects of climate change extend warming to include an increase in temperature variability and extreme weather events, but it is undeniable that temperatures are increasing, with the average global temperatures 1.76° F (0.98° C) higher in 2020 than the twentieth-century average, and 2.14° F (1.19° C) higher than the pre-industrial period (1850–1900).

A CHANGING SEA

The oceans are also warming, and simultaneously experiencing two additional major impacts of climate change: a reduction in dissolved oxygen, and an increase in acidity. However, it's much easier for research to tackle one factor at a time, so our scientific understanding of how marine life will be affected by this "triple threat" is limited.

The reason for reduced oxygen is complex. Partly it's because warm water can hold less dissolved oxygen than cool water. Partly it's because algal blooms are increasing, temporarily producing more oxygen, but ultimately dying and decomposing in a process that leads to a decrease in overall oxygen and the growth of existing oxygen minimum zones (pages 186–189).

As for the acidification, that's simpler—it's a direct result of carbon dioxide in the air dissolving in the water. The oceans absorb up to half the carbon dioxide in the atmosphere, acting as a significant buffer to climate change. This carbon dioxide reacts with water molecules to produce free hydrogen ions, which in turn react with the dissolved minerals that shell-building organisms use to make their shells, thus taking them out of circulation.

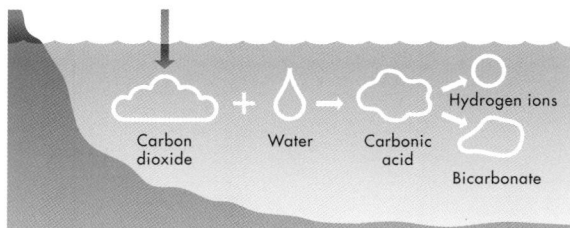

Carbon dioxide + Water → Carbonic acid ⇢ Hydrogen ions / Bicarbonate

Acidification
Carbon dioxide released into the air by human activity diffuses into the sea, reacting with water and carbonate to produce acid and make the building blocks of shells less available.

IMPACTS OF ACID

When we think of acid in the ocean, it's easy to imagine its destructive capacity to eat away at hard shells and skeletons. But the acidity of the ocean is not forecast to reach anything like the acidity of vinegar or lemon juice. It is possible for increasing ocean acidity to dissolve especially small or thin shells, like those of larvae; however, a bigger problem is the reduced availability of shell-building material. As it becomes harder to build their shells, animals won't be able to grow as fast or as large, or they might need to allocate energy toward growth that would otherwise have been used to protect them against disease or predation.

↑   This photograph of a storm arriving at an island in Micronesia recalls the importance of reefs as protective buffers. The loss of corals in our world ocean is not merely aesthetic.

TINY VITAL BONES

Statocysts are how cephalopods sense gravity and movement, determining up from down and gauging their acceleration. A cephalopod has two statocysts inside its head, each a tiny fluid-filled chamber lined with minuscule hairs. Inside this fluid is a calcified "stone" called a statolith. The statolith is heavy enough to sit on the bottom of the chamber, pressing on the hairs, which then send a nervous signal back to the brain. If the animal turns upside down, the statolith presses on the opposite side of the chamber. When the animal accelerates or decelerates, the statolith rolls forward or backward, changing which hairs it touches and giving the brain a new signal to interpret.

Scientists have studied the impact of ocean acidification on statolith development by rearing squid embryos at elevated concentrations of carbon dioxide, and found that acidic water causes their statoliths to grow smaller, in irregular shapes, and with more holes. If the statoliths are not formed properly, a cephalopod's movement and swimming are affected.

Statocyst

The balance organs called statocysts are not unique to cephalopods. Clams, crabs, and many other invertebrates have them too. Even we vertebrates possess a similar structure in our inner ear.

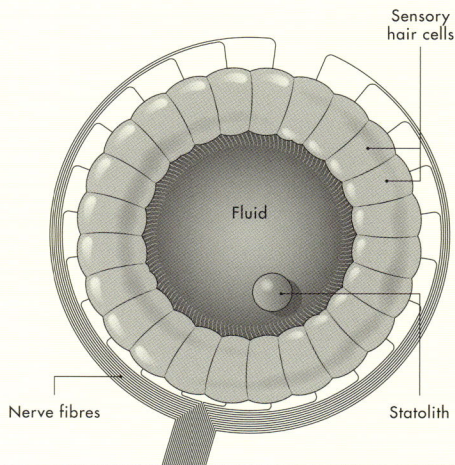

Sensory hair cells

Fluid

Nerve fibres

Statolith

Corals are clearly some of the most vulnerable marine organisms to acidification. Scientists have found that corals build thinner skeletons in more acidic waters, which leaves them at greater risk of breaking. Waves, storms, and animal activity that might not otherwise have damaged a reef are now a threat. In the Coral Triangle area of the Indo-Pacific, scientists have calculated that coral skeletons could lose up to 20 percent of their density by 2100.

Losses of corals will affect all the animals that depend on them, including cephalopods. Furthermore, these animals will also encounter impacts on their own bodies due to ocean acidification. How are cephalopods affected by acidification, bearing in mind that it is occurring in conjunction with increasing temperature and decreasing oxygen? The answer is not simple. Cephalopods that produce calcified shells face the same challenges as corals. That includes nautiluses, argonaut octopuses (*Argonauta* spp., pages 164–165), and cuttlefish. But even shell-less octopuses and squid depend on calcification for smaller, critical parts of their bodies: their statocysts, or "inner ear" equivalents.

STRUGGLING TO BREATHE

Acidification can also impact animals in ways that are not directly related to calcification. Hemocyanin, the molecule in cephalopod blood that binds oxygen the way hemoglobin does in our own, is extremely susceptible to changes in pH. Increased acidity affects its ability to bind oxygen, leading to a phenomenon known as respiratory acidosis. Cephalopods tend to be active animals that require a lot of oxygen, especially due to the inefficiency of jet propulsion. Studies on swimmers such as the Humboldt Squid (*Dosidicus gigas*) have shown that under more acidic conditions they may not be able to transport as much oxygen as called for by their metabolism. Warmer waters exacerbate this condition, since higher temperatures further increase metabolism, meaning the animal needs even more oxygen to maintain the same activity level.

Ocean acidification also affects growth, although the mechanism whereby this happens is not yet clear. Some studies suggest that embryonic and paralarval cephalopods grow more slowly and to smaller sizes under more acidic conditions. Other species, however, seem more tolerant—and many cephalopods could be among the most resilient marine life in the face of climate change. Octopuses and nautiluses are both quite capable of handling reduced oxygen availability by lowering their metabolic rate and, in the case of nautiluses, shutting themselves inside their shells and breathing oxygen from the gases in their chambers. A study that exposed octopuses to both increased acidity and decreased oxygen found that when first exposed to high acidity, the animals' need for oxygen increased, but if they had time to acclimate to the higher acidity, they were able to adjust their metabolism to cope with less available oxygen.

↑　Pteropods are small sea snails and sea slugs that spend their lives swimming near the ocean surface, using their feet as fins. The ones that grow shells are particularly susceptible to ocean acidification.

↖　When coral reefs are damaged by storms, bleaching, or other causes, the broken skeletons can be quickly overgrown by algae, making recovery far more difficult.

Communication on the reef

In the clear and well-lit water of coral reefs, visual signals evolve and diversify. Three animal groups have independently evolved a strong dependence on vision: vertebrates, cephalopods, and crustaceans. As we've seen, vertebrates and cephalopods evolved very similar eyes, while crustaceans (crabs, lobsters, shrimps, and all their relatives) have the compound eyes typical of other arthropods such as insects.

CRUSTACEAN COLORS

We may have gained a cartoonish impression that compound eyes see many copies of everything, but the arthropod brain integrates the images from many units, called ommatidia, into a single impression of their surroundings. Compound eyes can't resolve detail the same way that lens eyes can, but they can take in a nearly 360-degree continuous view of the environment. It's hard to sneak up on an animal with spherical eyes on stalks. The incredible camouflage of cephalopods is an invaluable asset when they're hunting crustaceans.

The pinnacle of crustacean vision undoubtedly occurs in the mantis shrimps, or stomatopods. These animals not only have good all-around vision, but also the greatest number of color-receptive pigments ever documented in the animal kingdom. We humans use our three cones,

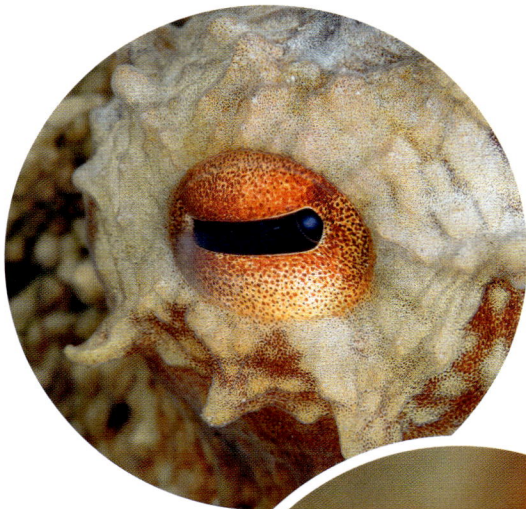

↖ The eye of an octopus and of a crustacean are extremely different yet both are extremely capable of perceiving the world. Light is focused through the single lens of an octopus eye (top), while each of the thousands of ommatidia of a crustacean eye has its own cornea, lens, and light receptors (bottom).

↗ In mantis shrimp eyes (left), a "midband" of specialized ommatidia separates an upper and lower hemisphere. Pseudopupils (black spots) are simply ommatidia that are pointed directly at the camera and reflecting no light toward it. The wide, curved pupil of this Bigfin Reef Squid (right) could potentially be used to shift the animal's vision across different wavelengths, offering a kind of color perception—a possibility which remains only theoretical for now.

which are sensitive to different wavelengths of light, to perceive colors. Mantis shrimps have 12 color receptors (wavelength-sensitive pigments), which has given rise to the popular concept of "shrimp colors," in which we imagine that these animals can perceive colors we humans can't. However, further research has revealed that mantis shrimp brains are not as capable as our own at comparing input from different receptors to obtain fine-grain distinctions between colors, so it's unlikely that their eyes can in fact perceive many more colors than our own.

Not that mantis shrimp eyes aren't still amazing. They can detect ultraviolet light (which we can't) and they can even detect circularly polarized light, an extremely rare type of polarized light that bounces off the carapaces of other mantis shrimp. Detection of polarized light often seems to arise as a mechanism to recognize members of your own species, as in the case of fish and cephalopods that can see non-circular polarized light.

SQUID PATTERNS

Of all cephalopods, it is reef squid whose visual communication has been studied in the greatest detail. The genus *Sepioteuthis* has representatives in reefs all around the world; the Bigfin Reef Squid (*Sepioteuthis lessoniana*) is profiled in this chapter (pages 138–139). Behaviors, including color patterns and skin changes,

can be documented in an ethogram—a behavioral inventory. In 2017, a group of scientists created an ethogram for the Bigfin Reef Squid during mating behavior, and discovered that this species shares many distinctive patterns with the Southern Reef Squid (*Sepioteuthis australis*) and Common European Squid (*Loligo vulgaris*), whose ethograms had been documented earlier, but more than half of its body patterns are brand new.

Bigfin Reef Squid mix and match their patterns in different, repeatable, yet flexible combinations when they engage in different behaviors. Although the documentation occurred during mating, the researchers identified patterns that serve non-reproductive purposes as well. Eyespots, for example, are displayed by males after successful matings, and also used in displays meant to frighten off predators. It's not much of a stretch to think of these components as words in a language that can have different meanings in different contexts. Scientists acknowledge that describing cephalopod visual communication as a language is still speculative, but it is certainly a framework that opens the door to more research.

The value of venom

Octopuses all have venomous saliva, which they use to subdue their prey and to defend against predators. So do some squid and cuttlefish. These venoms are commonly made of molecules called serine proteases, which are also components of viper venom. These molecules break apart proteins, which isn't intrinsically a bad thing—we produce and use proteases all the time, to digest the proteins that we eat, for example. However, to have proteases injected via a bite into the skin, muscle, or blood where they do not belong can cause significant pain and damage. They create bleeding, both internal and external, and necrosis, or tissue death.

TETRODOTOXIN

It is in the tropical coral reef habitats that octopuses have evolved their most potent venom constituent: tetrodotoxin, a chemical that affects the nervous system. As a category, neurotoxins have evolved in a variety of venomous animals, including non-viper snakes such as cobras and mambas. Tetrodotoxin itself has a more limited but still diverse distribution—it's not found in snakes, but in several species of fish, snails, worms, amphibians, and even a starfish. And, of course, this includes blue-ringed octopuses (*Hapalochlaena* spp.).

Tetrodotoxin specifically acts by blocking neuron communication. This produces paralysis when nerves cannot tell muscles what to do, and can lead to death. Fatal tetrodotoxin poisoning has occurred in humans both from eating pufferfish and from bites by blue-ringed octopuses, through failure of the diaphragm to maintain breathing. There's no antidote, but if a person can be kept alive by managing symptoms until the toxin is flushed out of the body, there are no known long-term effects.

SAFETY AND WARNING

Even a tiny amount of tetrodotoxin can be fatal, which leads us to wonder how the animals that produce it keep themselves safe. The genome of the Southern

Blue-ringed Octopus (*Hapalochlaena maculosa*) was sequenced in 2020, and scientists were able to use it to understand more about how these octopuses handle their toxins without being damaged by them. They found a "resistance mutation" that was familiar from pufferfish genes, which change nerve cells to make them less susceptible to the effects of tetrodotoxin.

The study also investigated the current hypothesis that tetrodotoxin is produced by symbiotic bacteria. The scientists discovered many different kinds of bacteria in the salivary glands of blue-ringed octopuses, but unfortunately most of them cannot be cultured in the lab. Thus, although a toxin-producing species has yet to be identified, at least we have a sense of the range of potential candidates.

Blue-ringed octopuses signal their toxicity with the vivid rings that give them their name, as do the Poison Ocellate Octopuses (*Amphioctopus mototi*, pages 136–137). This type of pattern is called aposematism, an honest signal to predators that the prey will be "unprofitable" (a polite way of saying "murderous").

↑ The muscle tissue of a Flamboyant Cuttlefish is highly toxic, and its colors may warn predators away. The patterns are also used in courtship.

← Instead of covering the blue iridophores of their rings with chromatophores, blue-ringed octopuses cover their rings with pouches of skin that can be expanded or contracted to hide or reveal them.

Poison Ocellates have an additional display of bright high-contrast stripes, also seen in the Ornate Octopus (*Callistoctopus ornatus*, pages 134–135), the Wunderpus (*Wunderpus photogenicus*, pages 70–71), and many other species—this is called a deimatic display. (Eyespots can contribute to a deimatic display as well, though they can also be seen as a form of mimicry.) Deimatic displays deter predators, but without conveying any information. They simply take advantage of the fact that most visual predators have a startle response. It seems fairly common for toxic animals to combine predator-deterrent techniques.

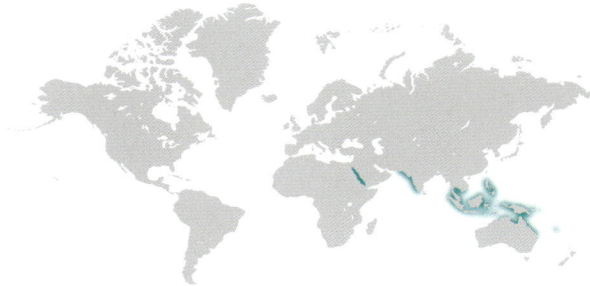

Star-sucker Pygmy Octopus

World's tiniest octopus

SCIENTIFIC NAME	*Octopus wolfi*
FAMILY	Octopodidae
MANTLE LENGTH	⅝ in (1.5 cm)
TOTAL LENGTH	1¼ in (3 cm)
NOTABLE ANATOMY	Frilly suckers on male arm tips
MEMORABLE BEHAVIOR	Large percentage of life spent in the planktonic stage

This species is the smallest octopus at adult size ever described. Living in and around corals from the western Pacific to the Red Sea, in many ways it's a typical octopus. Star-sucker Pygmy Octopuses hide from predators in holes and crevices and hunt as active predators themselves on tiny crabs, clams, and shrimp. Females lay a clutch of eggs, which they care for, and the hatchlings are planktonic paralarvae that drift and feed in the currents before settling as benthic adults.

However, this species has a slightly different take on each aspect of its life. These octopuses spend a surprising amount of time in the planktonic stage—three to six months. The animals live for less than a year, so that's a significant percentage of their life span. Additionally, females can lay two or three broods of eggs before dying, rather than only one. The rims of cephalopod suckers are typically grooved to enhance their ability to grip on uneven surfaces, but the grooves are microscopic. Star-sucker Pygmy Octopuses,

by contrast, go beyond grooves, but only on the suckers near the arm tips in males. The rims of these suckers are frilled, with projections like the rays of a star. The frills are called papillae, the same name that describes the bumps and projections elsewhere on octopus skin, although the sucker papillae cannot be changed in the same way. These star-suckers appear to maintain their function as ordinary suckers.

Although distinctive, this modification is not unique—at least one other species, *Octopus balboai*, also has papillae on the suckers near its arm tips. Other octopuses have unusual sucker modifications, including suckers that grow into long "tubercles" or "fingers." Such finger-like modified suckers are present in male Lesser Pacific Striped Octopuses (*O. chierchiae*, pages 100–101) and *O. penicillifer*. It always seems to be the males of the species that have these curiously shaped suckers, which suggests that they serve some sexual purpose, but what that could be has yet to be identified.

Although its coral habitat may render *O. wolfi* subject to the same risks that face corals and their dependent species everywhere, it is not currently considered a species of concern. It's too small to fish for human consumption, and has a varied diet that's not particularly threatened.

Tiny octopus

Octopus wolfi is the smallest of several dwarf octopus species, including the Atlantic Pygmy Octopus, Fitch's Lilliput Octopus, and the California Lilliput Octopus (which we met on page 83).

Octopus at actual size (25 mm)

→ The minuscule size of this octopus presents a particular risk in captivity. In tanks with bubblers, a single air bubble can become trapped in the animal's mantle and prevent it from breathing.

CALLISTOCTOPUS ORNATUS

Ornate Octopus

Nocturnal cannibal

SCIENTIFIC NAME	*Callistoctopus ornatus*
FAMILY	Octopodidae
MANTLE LENGTH	5 in (13 cm)
TOTAL LENGTH	3¼ ft (1 m)
NOTABLE ANATOMY	White stripes along mantle
MEMORABLE BEHAVIOR	Long-distance dispersal by paralarvae

Edible to humans, as octopuses tend to be, this species is fished for subsistence in its home range of the Indo-Pacific, but a commercial fishery has never been started. It's sometimes called the Night Octopus, in contrast to the Day Octopus (*Octopus cyanea*), which overlaps with its range and is of a similar size, but has an opposite activity cycle.

Ornate Octopuses are extremely fecund, laying egg clutches that contain up to 35,000 eggs. These eggs hatch as paralarvae that drift with the plankton and engage in long-distance dispersal. Planktonic larvae are a little bit like dandelion seeds that can be carried long distances on the wind. Ornate Octopuses can move much farther as tiny babies carried by currents than they can as benthic crawling adults, and this spreads the species over a larger geographic range, making them more resilient to environmental disturbances. Thus,

although Ornate Octopuses are well-known occupants of the Great Barrier Reef, they also live throughout the Pacific and Indian Oceans all the way to South Africa.

The name *ornatus* refers to the species' constant skin decoration of white spots and stripes, sometimes made even more noticeable by papillae lifting the white patches of skin. These highly noticeable stripes are not meant to be camouflage, but rather a deimatic display: an attention-grabbing pattern meant to frighten away predators.

Although many species of octopuses will attack and eat other octopuses, Ornate Octopuses are particularly well known for their cannibalism. Octopus beaks are often found in their stomachs, and members of other octopus species have been shown to avoid them assiduously.

Like several other octopus species—Wunderpus (*Wunderpus photogenicus*, pages 70–71), and the Atlantic Longarm Octopus (*Macrotritopus defilippi*, pages 68–69)—Ornate Octopuses have extremely long arms relative to their body. The uppermost pair of arms is longer and thicker than the others.

→ While all the arms of the Ornate Octopus are quite long, those on the upper or dorsal side are more muscular. This difference may affect their swimming and hunting in ways we don't yet know.

AMPHIOCTOPUS MOTOTI

Poison Ocellate Octopus

Iridescent eye-faker

SCIENTIFIC NAME	*Amphioctopus mototi*
FAMILY	Octopodidae
MANTLE LENGTH	4 in (10 cm)
TOTAL LENGTH	10 in (25 cm)
NOTABLE ANATOMY	Paired blue ocelli
MEMORABLE BEHAVIOR	Hunting with powerful venom

This species is a driller that uses venom to paralyze its prey. The radula, that tongue-like organ covered with small teeth common to many mollusks, can drill a hole right through a thick seashell. Unlike predatory snails, whose radula drill holes are large enough to slurp out their prey, octopus drill holes are too small for this, and serve only for venom insertion. Whether there's a clam or a snail or a hermit crab inside, the venom prevents its muscles from holding its body inside the shell, and the octopus can easily pull it out through the shell's opening.

Like another venomous coral reef cephalopod, the Poison Ocellate Octopus uses vivid blue rings to advertise its toxic nature. However, whereas blue-ringed octopuses are covered with these displays, the Poison Ocellate Octopus has only two: one on each side of its head, underneath its eyes. These blue rings are part of a larger high-contrast skin display called ocelli, or eyespots, with a dark circle inside a light circle.

These give the species it common name. Ocelli are considered a type of mimicry, looking like eyes but usually much larger than the animal's real eyes, with a greater distance between them and located in a different place on the body. The larger size and wider positioning may suggest a larger animal, making predators think twice before tackling it. They may also distract attention from the animal's real eyes, which are a vulnerable target.

The scientific name *mototi* comes from the word for "poison" in the Rapanui language of Rapa Nui (Easter Island). It may be more accurate to translate "mototi" as "venom," since scientific English distinguishes between poisonous animals, which are dangerous to eat, such as the poison dart frog, and venomous animals, which can inject toxin through a sting or bite, such as bees or vipers. Despite this technical linguistic distinction, the name "Poison Ocellate Octopus" has been used and published enough that it is considered the species' common name.

Similar to the fact that there are four distinct species of blue-ringed octopuses, there are also numerous octopus species with blue rings in their ocelli. The genus *Amphioctopus* contains several ocelli-bearing species, though the exact number is still under study. The chemical and molecular properties of their venom also remain mostly mysterious, since it has not been studied as extensively as the tetrodotoxin-based venom of blue-ringed octopuses.

→ In the same genus as the Coconut Octopus (*Amphioctopus marginatus*), the Poison Ocellate Octopus has similar long arms and hunting habits.

SEPIOTEUTHIS LESSONIANA

Bigfin Reef Squid

Cuttlefish squid

SCIENTIFIC NAME	*Sepioteuthis lessoniana*
FAMILY	Loliginidae
MANTLE LENGTH	10 in (25 cm)
TOTAL LENGTH	13 in (33 cm)
NOTABLE ANATOMY	Fins extending the length of mantle
MEMORABLE BEHAVIOR	Distinctive mating displays

The squid genus *Sepioteuthis* has evolved convergently with cuttlefish to possess long fins that can be used for slow but highly controlled swimming. (Hence its scientific name, which means "cuttlefish squid.") This is advantageous for life in a complex habitat with abundant hard structures, such as coral, which would be extremely damaging in even mild collisions. The fins allow these animals to hover in place as they hunt their prey: small reef fish and crustaceans.

Like other loliginid squids, Bigfin Reef Squid grow extremely quickly and have short life spans, well under a year. Humans have put this species to many uses, from fishing them for food to culturing them for display in aquariums. The squid's ability to control its swimming in reefs also serves it well in captivity, making it less likely than other species to bump into the walls of its enclosure (a situation that can lead to skin abrasions, infections, and lesions).

Bigfin Reef Squid are found throughout tropical and subtropical oceans from the Pacific and Indian to the Red Sea and even, recently, in the Mediterranean, where they are believed to have traveled by traversing the Suez Canal. Careful analysis suggests that "Bigfin Reef Squid" in different parts of the world may actually belong to different species, although they have not yet been separated and named.

Reef squid swim together in groups of mixed-size individuals, a behavior that's extremely unusual among squid due to most species' propensity for larger individuals to eat those that are smaller. Such cannibalism has never been documented in reef squid. Instead, they seem to communicate extensively with each other, to the point that some scientists have proposed the existence of a visual language in the closely related species the Caribbean Reef Squid (*Sepioteuthis sepioidea*).

The best documented patterns of Bigfin Reef Squid pertain to mating, although their exact meaning and purpose remain unknown. Both sexes were shown to exhibit two species-specific behaviors in captivity: "accentuated gonads," in which the animal makes the white ovary or testis show through the translucent mantle (nearby animals often respond with the same behavior), and "spread arms," which goes with a darkened mantle and pursuit of another squid, perhaps due to sexual interest or aggression.

→   Reef squid are known for swimming in groups, which some scientists suggest calling squads to distinguish them from fish schools. Bigfin Reef Squid begin swimming parallel to each other in a belt-shaped squad 1–2 months after hatching.

SEPIA BANDENSIS

Dwarf Cuttlefish

Smallest cuttle

SCIENTIFIC NAME	*Sepia bandensis*
FAMILY	Sepiidae
MANTLE LENGTH	2¾ in (7 cm)
TOTAL LENGTH	4 in (10 cm)
NOTABLE ANATOMY	Flaps on the underside of mantle
MEMORABLE BEHAVIOR	Walking on mantle flaps

This tiny cuttlefish has a funnel for jet propulsion and fins for maneuvering, but it also has a fascinating adaptation for walking: two flaps that protrude from the underside of its mantle, and which can be controlled (most likely with the same muscles that create papillae) in tandem with its lower arms, to walk along the seafloor.

Dwarf Cuttlefish can be kept and bred in captivity, where they will display all their typical behaviors from life in the wild: hunting prey, camouflaging with the background, eating each other if they don't get enough food, competing for mates. "You could do a reality show on cuttlefish," says Richard Ross, who cultured them at the California Academy of Sciences. Female Dwarf Cuttlefish produce eggs wrapped in a layer of inky jelly, which might be a deterrent for predators, pathogens, or both. But researchers have found that it is possible to remove this layer and raise the embryos without their jelly to get a clear view of their development. They hatch not as planktonic paralarvae but as miniature adults, which are far easier for aquarists to feed.

Because of the relative ease of keeping them in captivity, Dwarf Cuttlefish are also the subjects of many areas of ongoing research. In 2020, young Dwarf Cuttlefish displayed signs of learning and memory when scientists tested their ability to change their hunting behavior. The researchers presented the cuttlefish with prey inside a transparent tube, so that it was visible but inaccessible. The cuttlefish learned not to strike at this unavailable food. An earlier study with the Common Cuttlefish (*Sepia officinalis*) had shown that juveniles, while able to learn in the short term, did not retain the information over multiple days. Juvenile Dwarf Cuttlefish, by contrast, demonstrated long-term memory of what they learned. The authors postulate that Dwarf Cuttlefish brains develop faster to facilitate visual learning because of their habitat—a complex reef in clear tropical water. (By contrast, Common Cuttlefish grow up in a relatively low-visibility temperate habitat.)

This theory may be easier to test now that we have a brain atlas for this species. In 2022, a group of scientists put Dwarf Cuttlefish through an MRI machine to scan and map their brain anatomy and activity. The result is a resource that will facilitate future work on the connections between behavior and neurons.

→ Cuttlefish of many species are often observed raising and spreading their arms. This behavior can help with camouflage, or it can be a threatening posture to increase the animal's apparent size.

NAUTILUS MACROMPHALUS

Bellybutton Nautilus

"Innie" shell

SCIENTIFIC NAME	*Nautilus macromphalus*
FAMILY	Nautilidae
SHELL DIAMETER	6¼ in (16 cm)
NOTABLE ANATOMY	An open shell coil
MEMORABLE BEHAVIOR	Nightly presence near surface

The adorable name of this species refers to the innermost coil of its shell. Nautiluses begin growing their shells while still in the egg, building each new chamber on the previous one. The shell starts off very small and becomes larger and wider as the animal grows and coils around itself. This creates a cone-shaped space called the umbilicus leading from the outermost whorl into the center of the shell. In other species of nautiluses, the umbilicus is covered by a callus, but not in the Bellybutton Nautilus. The reason for this remains a mystery—like so many other things about nautiluses.

The Bellybutton Nautilus is also the smallest species of nautilus. Its range is limited to New Caledonia where, during the night, it also gains the distinction of being one of the shallowest nautilus species. All nautiluses go through a diel migration, moving into deep water to hide from predators during the well-lit day, and rising to shallow waters to hunt at night. "Shallow" for other nautilus species still means 230–260 ft (70–80 m) below the surface. However, Bellybutton Nautiluses have been spotted by divers on reefs at less than 66 ft (20 m).

Observers have sometimes thought that nautiluses might accomplish their vertical migrations by increasing and decreasing the amount of gas in their shells. However, we know now that the process of moving gas into the chambers happens on a much slower timescale. An ingenious tube called a siphuncle, which is part of the nautilus's living body, goes all the way back through every chamber in the shell. The blood in the siphuncle does not come into direct contact with the fluid (seawater) in the chambers, but the membrane of skin between the two is permeable. The nautilus actually has control over the salinity of the blood in the siphuncle and, by making it extra salty, the animal causes water to be drawn out of the chambers and into the blood by osmosis. This creates a negative pressure in the chambers, which then draws dissolved gases out of the blood and out of solution, filling the chamber partially with gas.

For their daily migrations, nautiluses use their slow but reliable jet propulsion.

Open umbilicus

Most species of nautilus grow a calcified callus over their inner "baby coils" (A), but the Bellybutton Nautilus, as well as nautiluses of the genus *Allonautilus*, leave this umbilicus exposed (B).

A

B

→ Nautilus tentacles have no suckers. Sometimes called cirri, they can be extended from or retratcted into a thick sheath, and are covered with sensory receptors that we're still just beginning to understand.

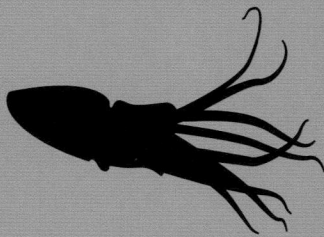

OPEN OCEAN

The sunlight zone

Past the continental slope and the continental shelf is the open ocean. Here, the seafloor drops away thousands of feet and is essentially irrelevant to the creatures living far above it. Later, we'll follow the ground from the slope and shelf into the deep abyssal plain (see pages 212–247), but for now, let's migrate horizontally, maintaining the depth of kelp forests and reefs to stay in the sunlit zone, the upper 660 ft (200 m) or less. What happens in these waters when there's no longer any land beneath them?

MICROSCOPIC ALGAE

The base of the open ocean ecosystem is plankton. Plankton is a general term that means "drifter" and describes any living organism that is primarily pushed around by waves and currents, rather than moving under its own power. In one sense, everything gets moved by the ocean. If you've ever gone swimming in the sea, you've probably been moved at least a little by a wave or a current. Taking into account storm surges and riptides, there are situations in which nearly every organism in the sea would count as plankton. But in its usual sense, plankton encompasses organisms that do not typically swim against even mild currents—and most plankton are tiny, including the algae that support the rest of the open ocean food web.

Kelps and corals, the great photosynthesizers of the continental shelf, cannot live out here where there is nothing to attach to. (With one notable exception— *Sargassum* spp., a seaweed that can grow in the plankton and lends its name to the Sargasso Sea.) Instead of large organisms, the dominant photosynthesizers of the open ocean are microscopic single-celled algae, such as the glass-shelled diatoms. They can live alone or aggregate into colonies, although even these linked colonies remain microscopic.

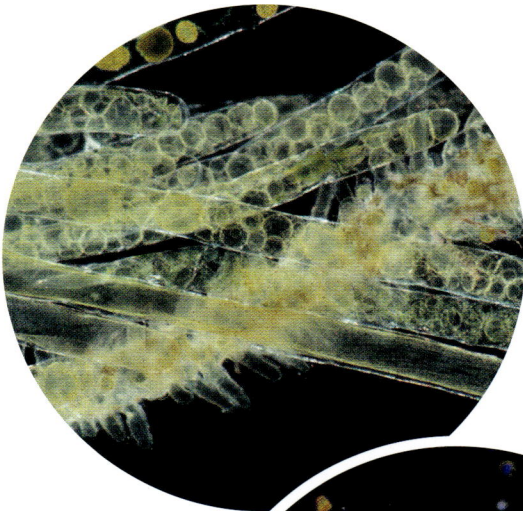

Small they may be, but their effects are enormous. These phytoplankton, or photosynthesizing plankton, produce up to half of Earth's oxygen. Like other algae and plants, they require carbon dioxide to power their metabolism, which drives a carbon sequestration process known as the "biological pump." Carbon sequestration is a topic that has become more familiar in recent years, as we continue to release more carbon dioxide into the atmosphere with troubling effects. Photosynthetic organisms such as trees and diatoms can offset this increase by taking up the carbon and building it into their bodies. Phytoplankton live such short lives that they quickly transport the carbon into the deep sea by dying and sinking. The rate of their sinking and the overall influence of the biological pump on climate change is an area of active research. Some would-be climate engineers have even proposed attempting to fertilize areas of the ocean to promote increased phytoplankton growth that would draw down even more carbon dioxide, but this experiment has not been proven feasible.

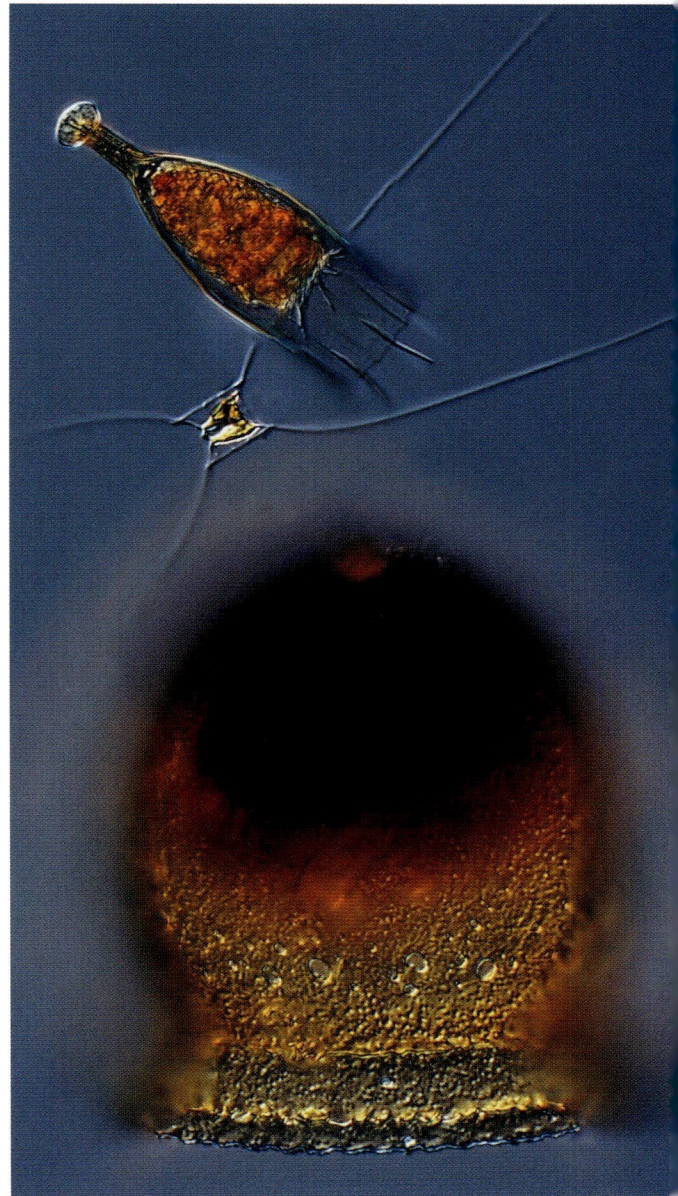

↑ Photosynthesizing diatoms create the essential fatty acids that accumulate in grazing zooplankton and in the fish that eat those plankton. They are therefore responsible for the health benefits of fish oil.

← Single-celled planktonic algae often grow in chains (top). They are eaten by tiny animal plankton, called zooplankton, such as the copepod pictured here (bottom).

INTRODUCING ZOOPLANKTON

The next link in the food webs supported by phytoplankton are the zooplankton that graze on the algae. The dominant planktonic herbivores are copepods, tiny zooplankton that resemble small shrimp and are sometimes called the "cows of the sea" because of their relentless grazing habits. Also like cows, they produce a substantial quantity of feces. These, too, sink from the sunlit waters to the depths and contribute to the biological pump.

In any ecosystem, cephalopods are reliable predators of crustaceans, and the open ocean is no exception. Which are the cephalopods in the planktonic food web that tackle copepods? Primarily it's paralarvae, the hatchlings that inhabit the plankton until they're old enough to either settle to the seafloor or develop into strong swimmers. As far as we know, paralarval cephalopods, like their parents, never engage in herbivory. However, the paralarvae of flying squid (Ommastrephidae) are detritovores, more likely to eat the fecal pellet of a copepod than the copepod itself.

↖ Transparent gelatinous animals called salps are common in the zooplankton, although they are capable of swimming as well as drifting. In fact, salps are the only animals besides cephalopods to engage in jet propulsion.

↑ Active swimmers like this adult reef squid are not considered part of the plankton, although their drifting paralarvae are.

↗ Predator–prey interactions among planktonic babies can be fierce. Here a juvenile Jewel Squid attacks a paralarval Fire Squid.

But most cephalopod paralarvae are predatory, with many meal options beyond copepods, including all the other babies in the plankton. Many lobsters and crabs have planktonic larvae, as do fish, creating a rich array for cephalopods to eat—and be eaten by. Perhaps for protection from such fellow predators, young pelagic cephalopods of several species (including argonauts, *Argonauta* spp., and Football Octopuses, *Ocythoe tuberculata,* pages 166–167) have been found sheltering inside the gelatinous barrels of salps. These are planktonic drifters that look like jellyfish but are more closely related to vertebrates. Male argonauts and Football Octopuses remain tiny throughout their lives, and continue their salp-hiding habit into adulthood, while females grow large enough to swim actively and exit the plankton.

PLANKTONIC ADULTS

Young cephalopods of other species may grow larger but remain planktonic throughout their lives. These passive drifters lead an existence that is more similar to that of a jellyfish, following the currents and collecting prey as they encounter it. It must be noted, though, that even jellyfish often drift across the fuzzy boundary of what constitutes plankton. They aren't strong swimmers like a squid or a shark, but if currents are weak, a large jellyfish can move under its own power.

An additional similarity between jellyfish and some cephalopods lies in their transparency. Living in the sunlit region of the sea, with no background to match and no rocks to hide under, there's only one real option for camouflage: to look like the water itself. Many open ocean cephalopods are astonishingly good at doing this. They have evolved a reduction in chromatophores, iridophores, and leucophores to enhance the transparency of their bodies. They use photophores, light-producing organs, to block the shadows of the few body parts that cannot be made transparent, such as eyes and digestive glands. Their arms are often much reduced in length and thickness to reduce their silhouette.

Migration

While the photosynthetic basis of the open ocean food chain is of necessity constrained to the sunlight zone, the grazers that consume this algae only need to be there when they're eating. They themselves are small prey items with abundant predators, and the more time they spend in well-lit water, the greater their risk of being eaten. Thus, a strategy has emerged that shapes much of life in the open ocean: diel migration.

A DAILY UP AND DOWN

"Diel" refers to something that happens on a 24-hour cycle. Some texts refer to diel migration as "diurnal" migration, but technically "diurnal" means "during the day." Diurnal is the opposite of nocturnal. Diel migration, however, does not happen exclusively during the day nor exclusively at night. In fact, the actual movement of animals happens at the boundaries: dusk and dawn.

The small animals that graze on algae have evolved the habit of eating at night, when surface waters are dark and they are less visible to predators. During the day, they sink down to dimmer depths. Although they depend on sunlight to grow their food, they rarely experience daylight for themselves. Swimming up and down in the water takes energy, but it's worth it for the survival boost.

← Sardines are plankton predators that disperse at night to hunt their prey in shallow water. During the day, when the plankton sink and danger grows, sardines aggregate into dense schools for safety.

A MOVING FOOD WEB

The deep scattering layer is the name for the mass of animals that sinks to depth during the day and rises toward the surface at night. It was so named because it is thick and solid enough to reflect or "scatter" the sonar that ships use to detect the depth of the seafloor beneath them. Navy sailors during World War II were the first to notice it, and although they knew that the seafloor itself could not be moving up and down, they speculated that perhaps enemy submarines were causing the unusual readings.

Researchers figured out that the layer was composed of marine animals migrating in dense aggregations and, in the decades since, the phenomenon has been studied in detail. It's described as the largest movement of animals on the planet, and it contributes significantly to the biological pump. Many different kinds of crustaceans, fish, and cephalopods are part of the deep scattering layer, but they are not evenly mixed nor exactly synchronized. Within the layer, they gather in single-species schools or shoals, each of which moves up and down on its own schedule. Stronger swimmers that have a greater ability to avoid predators move toward the surface earlier in the evening, while weaker swimmers wait for greater darkness.

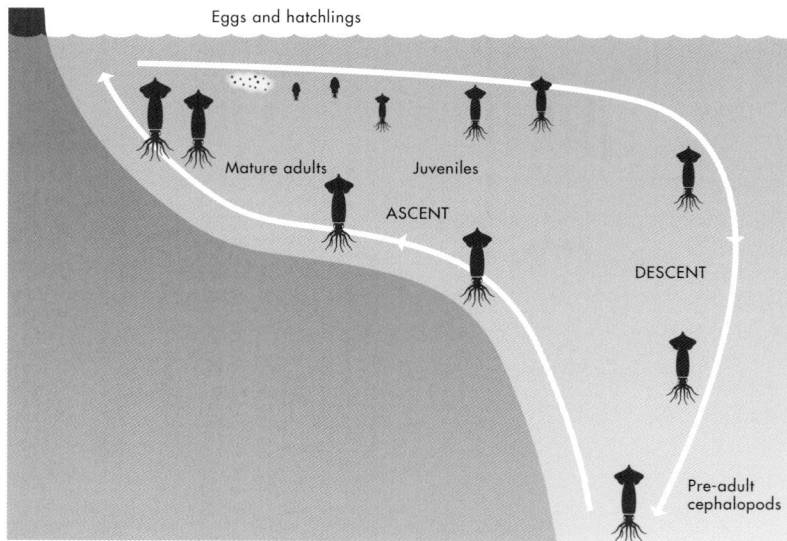

Eggs and hatchlings

Mature adults   Juveniles

ASCENT

DESCENT

Pre-adult
cephalopods

A 2021 study discovered that squid in the deep scattering layer demonstrated a particularly variable migratory behavior depending on predator risk. Dolphins can hunt at night using echolocation rather than vision, but they are tied to the surface by their need for air. So when dolphins are present, squid hide from them by remaining at depth for part of the night.

Water-breathing predators that are not tied to the surface, such as Humboldt Squid (*Dosidicus gigas*), will sometimes engage in a diel migration of their own to follow their prey in the deep scattering layer. Then, of course, the next level of predator becomes aware of and adapts to these habits. In other words, humans fish for Humboldt Squid at night when they get close to the surface.

DEVELOPMENTAL DESCENT

Humboldt Squid and many other cephalopods begin their lives as tiny vulnerable members of the plankton. At first, they cannot make significant diel migrations and remain mostly near the surface, but as they grow, they can sink to greater depths during the day and still manage the trek back to the surface at night.

Unlike Humboldt Squid, many cephalopods that

Ontogenetic migration
This refers to an animal's predictable changes in habitat over the course of its life. Many marine animals stay near the surface as larvae and move into deeper water as they mature, either returning to shallow water to spawn or releasing buoyant eggs that carry themselves into surface waters.

live near the surface as paralarvae descend to a more permanent life in the deep as adults. This phenomenon is called ontogenetic migration—movement from one habitat into another over the course of an animal's lifetime. Salmon and eels may be familiar examples. Eels are born in the ocean and migrate into rivers, while salmon are born in streams and migrate to the sea. For purely salt-water animals like cephalopods, ontogenetic migration from one marine habitat to another is common.

The open ocean octopus called an argonaut has a particularly fascinating method of planktonic movement that transitions as it ages. Very young argonauts, including recent hatchlings, are sometimes spotted clinging to the tops of jellyfish. These small octopuses are not yet strong swimmers, and they benefit from the swimming capacity (limited as it is) of the jellies. The predator deterrent of the stinging tentacles is probably also a benefit.

As the argonauts get larger, some have been found taking bites out of their rides, and eventually the argonauts grow to be stronger swimmers than jellies. Large argonauts still ride jellies at times, possibly continuing to benefit from the protection they offer as well as the gelatinous food source. They seem content to go at the pace of their pulsing mounts—perhaps it saves them the energetic cost of swimming.

Adult argonauts have clearly graduated from the plankton, as they can swim significant horizontal distances. Still, they illustrate the vulnerability of all marine animals to the caprices of the sea. Unusual conditions can lead them to be stranded on beaches, the same fate that can even doom some of the strongest of marine swimmers, whales.

HORIZONTAL MIGRATION

Whales are well known for engaging in long-distance horizontal migrations from their spawning grounds to their feeding grounds. Gray whales, for example, have the world's longest migration of over 10,000 miles (16,000 km) between summer feeding in the Arctic and winter calving in Mexico. Many cephalopods also engage in seasonal horizontal migrations, although the relative speed of their generational turnover and the complexity of their population dynamics makes it trickier to work out the details. Flying squid are the most well-studied seasonal migrators, with Humboldt Squid regularly crossing the Gulf of California, east to west and back again, and Japanese Flying Squid (*Todarodes pacificus*) moving north then south across the Sea of Japan.

→ A female argonaut with arm webs covering her shell rides on a jellyfish—or is she dragging the jellyfish along as a traveling snack? The exact nature of this relationship remains unclear.

Fishing on the high seas

The two cephalopod species most fished in the western Atlantic are the Longfin Squid (*Doryteuthis pealeii*) and the Shortfin Squid (*Illex illecebrosus*). They belong to the two different groups of squid: Myopsida and Oegopsida, which are often considered neritic and pelagic squid, respectively. The names of these groups, however, do not refer to their habitat preferences but to a fascinating anatomical distinction. Myopsida comes from the same root as "myopia" and refers to the eye-covering cornea that these squid possess. This cornea protects the lens and is sometimes confusingly called an eyelid. Oegopsid squids lack a cornea and their eyes are completely exposed to seawater; they have the ability to cover their eyes with true skin eyelids for protection.

ALONG THE CONTINENTAL MARGIN

Myopsid squids do tend to live near shore, while oegopsid squids are more likely to be found in the open ocean, but many species cross over this boundary or sit directly on top of it. Longfin Squid and Shortfin Squid both spend time over the continental shelf in the western Atlantic, with the myopsid Longfin Squid closer to shore and the oegopsid Shortfin Squid further out to sea. This illustrates the fact that the categories of "nearshore" and "open ocean" are not strictly obeyed by the animals we apply them to.

Both species are caught with bottom trawls, large open-mouthed nets dragged along the seafloor. As you might imagine, trawl nets are not species-specific, and will pick up anything and everything that fits in their mouths. United States fishing regulations state that in order to fish for Shortfin Squid, trawling must occur deeper than 50 fathoms (90 m)—not because Shortfin Squid don't swim any shallower, but to avoid catching Longfin Squid.

To reduce the impact of the trawl nets on whales and dolphins, fishers are required to follow guidance from boats and planes that look out for "marine mammal hotspots" and then direct fishers away from locations of high activity where these animals would be at risk of entanglement. Bycatch of non-mammal species is far more common, and has to be reported, even for species that are deliberately fished themselves, such as Atlantic butterfish. It doesn't matter for the health of the population if butterfish are caught by accident or on purpose; managers must keep track of all the numbers and close the squid fishery if butterfish bycatch gets too high.

↗ Atlantic butterfish, sometimes caught as bycatch in the squid fishery, are carefully managed by the U.S. government.

← Since many large marine species have been overexploited, we have started eating smaller ones, such as Longfin Squid, which were originally used for bait but have been sold for meat since the 1960s. This trend is known as "Fishing down the food web."

Because of the attention paid to bycatch as well as our knowledge of the squid populations themselves, both Shortfin and Longfin Squid are considered "good seafood choices" by the National Oceanic and Atmospheric Administration. In the case of Shortfin Squid (which is representative of most pelagic squid species around the world) we don't know how big the population is, while for Longfin Squid (which lays its eggs on the seafloor in predictable places and is easier to track), managers have an estimate of the stock size. Both species fluctuate significantly in numbers from year to year, so the managers set a different quota each season based on estimates of how many squid are available. So far, both fisheries seem to be managed at sustainable levels.

spawning grounds, Japanese Flying Squid are targeted on their feeding grounds. When catches of this species dropped in the 1970s, concerns of overfishing were raised, but the numbers climbed again in the 1990s, and research suggests that the stock sizes may be more strongly affected by environmental conditions than by fishing pressure.

ILLEGAL FISHING AND GHOST GEAR

Some squid species are likely targeted by IUU, or Illegal, Unreported, and Unregulated fishing. This category is a bit like poaching on land, except it's easier to get away with in the ocean. The high seas are not under the jurisdiction of any one country, but that doesn't mean there are no rules—international agreements have placed regulations around fishing these waters, since they are considered a shared resource. It's just extremely difficult to enforce them. IUU hurts individual fished species, the overall oceanic ecosystem, and many people as well. Often the people working on these fishing boats are tricked on board and suffer inhumane conditions to pay off debts, with all of the profits going to boat owners.

When fishing boats don't report their catches, their impacts on the environment cannot be tallied or understood. These impacts include the animals caught on purpose, the bycatch caught and discarded, and the litter of old fishing gear. Derelict fishing equipment includes lines that are cut and left in the sea, nets that are lost or discarded, traps that are never brought up, and more. This "ghost gear" can be created by legal fishing operations as well as illegal ones, and there's very little being done about it, although it can keep on killing marine life as long as it's in the sea.

OPEN OCEAN FISHING

Japanese Flying Squid (*Todarodes pacificus*) are not caught by trawl, but either by jig and line (like Humboldt Squid, *Dosidicus gigas*) or by different kinds of nets that go through the middle of the water, including gill nets. A gill net is like a wall in the water, with weights holding it to the seafloor and floats holding it at the surface. Fish or squid (or seabirds or mammals) swim into the net and become trapped and unable to escape. In the case of fish, the material of the net often catches behind their gill covers, hence the name of the net. Gill nets, like trawls, can catch a lot of things they don't mean to.

Migrations of Japanese Flying Squid between their feeding grounds and spawning grounds have been closely studied, since this species is commercially the most important cephalopod fished around Japan and Korea. Scientists have identified two stocks: one that spawns in the fall and one that spawns in the winter. Unlike Market Squid (*Doryteuthis opalescens*), which are fished on their

↖ The global catch of Japanese Flying Squid peaked in 1996 at over 700,000 tons. It has since declined to just over 100,000 tons in 2020, possibly due to overfishing.

↗ Illegal fishing includes a wide range of activities, from these trawlers operating in a protected marine reserve to the use of outlawed fishing methods or falsified reports.

→ Albatrosses use their sense of smell to seek out marine prey as they soar over the seas. They often scavenge dead floating cephalopods.

ALBATROSS PREDATORS

One aspect of squid fisheries that has yet to be studied in depth is the impact on marine predators of squid. Squid are favored prey of many fish, mammals, and seabirds—especially albatrosses. But albatrosses are not diving birds, so how do they get to their prey? Scientists have theorized that they eat only squid that have already died and floated to the surface—open-ocean squid are some of the few animals that float rather than sink after death, because of the ammonia in their tissues. It may be that the very existence of numerous albatross species has been facilitated by the spawn-and-die lifestyle of oceanic squid.

Plastic pollution

Ghost gear is one kind of open ocean pollution. But the prevalence of pollution in the sea goes far beyond lost nets. Plastic is the most significant factor, comprising 80 percent of marine debris, with rough calculations that the annual plastic input to the ocean is 14 million tons. It's amazing how far plastic can travel, reaching every part of the sea no matter how deep or how distant from land.

HOW PLASTIC GETS INCORPORATED INTO MARINE LIFE

Plastic of different colors and sizes resembles the food of different marine animals. Seabirds die because their guts are full of plastic; sea turtles asphyxiate when they try to eat plastic bags. Cephalopods seem less prone to eating plastic, perhaps because it doesn't look as much like their prey. Their short life span is also advantageous here, as each individual cephalopod doesn't have much time for plastic to build up in its body and cause problems before it's spawning the next generation and bowing off the stage.

Plastic goes beyond a dietary risk. It changes and shapes the environment. Once plastic is drifting at sea, it becomes a habitat for animals to attach to and live on. Similar to coastal octopuses crawling inside beer bottles, pelagic cephalopods and other animals may be drawn to plastic materials. Removing the litter becomes a complex calculation. We don't want to thoughtlessly collect all the beer bottles and doom the octopuses inside; similarly, collecting plastic from the ocean without also collecting and killing the animals living on and near it is a challenge that will require a great deal of research and finesse.

MICROPLASTICS

Plastic is broken into small pieces by weathering, like rocks breaking down into sand. It might be tempting to think that once plastic pieces are the size of sand grains, they're not a problem anymore. Unfortunately, microscopic pieces of plastic are under increasing scrutiny as a risk to all life on the planet, including humans. The physical deconstruction of an object,

↑   The term "microplastic" was coined in 2004 to describe fragments of plastic smaller than five millimeters, which have since been found in every environment on Earth. The even smaller "nanoplastics," measuring less than one micrometer, are likely as ubiquitous, but incredibly difficult to study.

←   Effective interventions for the problem of plastic at sea include simple contraptions that catch trash in rivers before it flows to the sea, and coastal cleanup efforts that remove garbage from the shoreline.

from a rock into sand or from a plastic fork into microplastics, does not change its chemical composition, and plastic chemicals can cause a variety of concerning health problems.

Living organisms, which evolved in an environment full of rocks and sand, can break them down and use their constituent minerals—salt, calcium, iron. However, living organisms did not evolve to handle plastic. No matter how tiny, plastic cannot be digested and turned into energy or body-building materials … yet. Scientists have identified a few bacteria and insects that may be able to digest plastics in small quantities, but we're still far from harnessing this ability to clean up the environment.

Research on the high seas

Although humans have been traveling across the open ocean for longer than recorded history, active research in this area is a relatively recent phenomenon. Early voyagers no doubt encountered flying squid, but it's unlikely that they sought remote squid fishing grounds, or spent time dragging nets through the water to examine what might be living deeper. Cephalopods in the open ocean were reported when they were exceptionally large and visible, like giant squid vomited up by whales at the sea surface, but it was scientific expeditions in the 1800s that began the process of deliberately searching for, cataloging, and giving scientific names to the creatures out there.

CLINES AND EGGS

The density of water in the open ocean is variable. Some regions are saltier and therefore more dense; some areas have a lot of freshwater influx and are therefore less dense. Cold water is denser than warm water. If different masses of water are not agitated with enough movement to mix them, then a blob of dense water will simply sink below a blob of less dense water. This happens all around the world, producing pycnoclines, or layers where density abruptly increases as you move deeper. (These often occur in conjunction with thermoclines, places where the temperature abruptly drops as well.)

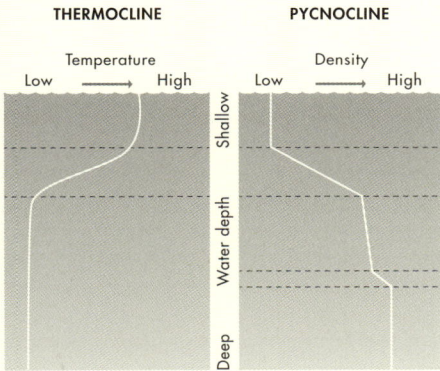

THERMOCLINE	PYCNOCLINE

Thermoclines and Pycnoclines

Thermoclines are created when warm water, which is less dense than cold water, sits in a layer on top of it. Temperature affects density, so pycnoclines or shifts in density often co-occur with thermoclines (upper dotted lines). Pycnoclines can also occur independently, due to other factors like salinity (lower dotted lines).

← Early "research" tools were simply fishing tools, like the harpoon sticking out of this enormous squid.

↓ Swimmers and snorkellers may encounter the egg masses of Diamondback Squid, due to the buoyancy of the gelatinous material, which causes them to float just below the surface.

The various parts of living organisms also vary in density, which in turn affects their buoyancy. For example, low-density gas inside a nautilus's shells offsets the high density of the shell material. The egg mass of the Diamondback Squid (*Thysanoteuthis rhombus*) has low-enough density to float just under the sea surface. Thus it is the egg mass most commonly observed by humans, because it can be spotted from a boat without needing to drop a net or a diver. Why does this egg mass float when others do not?

Flying squid egg masses have been far more difficult to find, with egg masses of most species never having been observed in the wild. From laboratory studies, scientists determined that these egg masses are denser than surface waters, and likely sink until they reach a pycnocline where they can rest above water with greater density than the mass itself. The first Humboldt Squid egg mass discovered did indeed sit on a pycnocline.

We can guess that this is advantageous for the developing embryos, keeping them away from surface waters where the environment is more stressful. Ultraviolet radiation from the sun can damage cells, wave and wind action could break apart the mass, and of course the embryos and hatchlings would be more visible to surface-based predators. So why don't Diamondback Squid make denser egg masses? We still don't know.

BUOYANT SHELLS

Speaking of buoyancy, it's time to consider a couple of cephalopods with unusual shells. The first, the argonaut octopus, has a shell that superficially resembles a nautilus, and has even been called the "paper nautilus" for the thin and fragile nature of its shell. However, its shell is only truly fragile when taken out of water and dried (which is how most of the European scientists who named things encountered it). In their natural watery environment, argonaut shells are fairly sturdy and flexible, although thin and translucent. Early observers suggested that argonauts sailed at the surface of the water, using their shell as a boat and two of their arms that bear expansive membranes as sails. In the nineteenth century, these membranes were proven to be used instead for the purpose of creating the shell.

A century and a half later, scientists discovered that argonauts can use their shells as a kind of buoyancy-controlled submarine. At the sea surface they get a bubble of air inside the shell, then swim down with it. As air is brought deeper, it is compressed by the water pressure all around. The air bubble shrinks and its buoyancy decreases until it exactly balances out the weight of the argonaut, and the argonaut with its air bubble has achieved neutral buoyancy. It has to expend no energy now to maintain itself at depth, and can devote all its swimming effort to horizontal movement.

Finally, there's an entire group of cephalopods with a delicate coiled internal shell—as thin as the shell of an argonaut, as coiled as the shell of a nautilus, as internal as the shell of a cuttlefish. These are the spirulids, and although they were once a speciose group, we know this diversity only from fossils. A single species remains in today's oceans: *Spirula spirula*, the Ram's Horn Squid (pages 172–173). It hangs vertically in the water and engages in diel migration. It shares this habit with many other open ocean squid, as it also shares with them the capacity for bioluminescence.

LUMINOUS LIFE

We've already met some bioluminescent cephalopods in the form of the bobtail squid and the Luminous Bay Squid (*Uroteuthis noctiluca*, pages 112–113), but there are nearly as many ways to be bioluminescent as there are ways to be a squid. (Squid scientist Sarah McAnulty has proposed "there are a hundred ways to be a squid" as a kinder replacement for the saying "there's more than one way to skin a cat.") Some open ocean squid, such as the Ram's Horn Squid, have a single light organ. This light organ is on the tip of its mantle, which puzzled scientists who assumed that the mantle tip would point upward, and therefore the light organ would be on the wrong side for counterillumination. Then in 2020 the first video of a Ram's Horn Squid in the wild revealed that they actually swim with the tip of the mantle pointed down, placing the light organ ideally for counterillumination. Other open ocean squid have multiple light organs. Some are on the tips of the arms. Others are studded all over the body, such as those found on jewel squid (*Histioteuthis* spp.). Are these light organs fueled by symbiotic bacteria, or do these cephalopods have their own way to produce light? We don't know. What purposes do these light organs serve? Photophores, such as chromatophores and iridophores, could be used to communicate between members of the same species—to attract a mate or to compete. But this is mere speculation.

Sometimes we think of bioluminescence as limited to the deep sea, where sunlight never reaches. But even the sea surface is dark for half the day (on average) and in this darkness, luminous animals shine.

←   This cross section through a nautilus shell shows the chambers that make cephalopod shells unique. Most of the animal's body is in the outermost living chamber, but a thin tube of flesh extends through all the chambers, modulating the amount of gas and fluid inside and thus controlling buoyancy.

→   The Ram's Horn Squid protects itself with color-changing skin and a counterilluminating photophore on the tip of its mantle.

ARGONAUTA NODOSUS

Knobbed Argonaut

Shelled octopus

SCIENTIFIC NAME	*Argonauta nodosus*
FAMILY	Argonautidae
MANTLE LENGTH	4 in (10 cm)
TOTAL LENGTH	12 in (30 cm)
NOTABLE ANATOMY	Egg case with bumpy ridges
MEMORABLE BEHAVIOR	Periodic stranding

Argonauts are unique among octopuses, among cephalopods, even among mollusks. They produce a shell, but it is grown only by the female, and it is grown not with her mantle but with her arms. It is used as an egg case in which she lays and broods her babies, but she begins to grow it at birth and remains in it throughout her life. She is not physically attached to it, since she can crawl in and out, but she is behaviorally attached to it, since she will not willingly leave nor does she seem able to make a new one from scratch as an adult. However, she does know how to mend it, and can fix cracks and breaks not only by secreting new shell material, but by picking up old shell pieces and gluing them into position.

→ This species was originally named *Argonauta nodosa*. This seems to follow the rules of scientific nomenclature, which require species and genus to agree in gender, as –a is typically a feminine ending. However, "nauta" is a masculine word for sailor, and thus the species had to be revised to the masculine "nodosus."

Argonaut shells are often found in the absence of the living animals, and their shapes can vary significantly based on each individual's lifetime experience of damage and repair. This has led to confusion over the number of species. Currently, only four distinct species of argonauts are recognized. All four produce shells with regular ridges, but only the Knobbed Argonaut produces knobs or bumps along these ridges, the habit for which it is named.

Male argonauts make no shell and grow to a fraction the size of the females, no more than 1½ in (4 cm) in total length. Their hectocotylus or reproductive arm, used to transfer spermatophores, grows nearly the size of their body, and detaches at the time of mating. There is no evidence that they can regenerate a new one. Due to their small size, they are observed much less frequently than females.

Mass strandings of Knobbed Argonauts are known from all coasts lining the oceans they inhabit, from Australia to Africa to South America. In some cases, these strandings have been clearly caused by ocean conditions, such as masses of oceanic water moving toward the shore and mixing with unusually warm waters, but in other cases the cause of stranding remains unknown.

The risk of ocean acidification to Knobbed Argonaut shells has been studied by soaking shells in water of varying acidity. Scientists demonstrated that the pH values projected for both 2070 and 2100 produced etching and dissolution of the argonaut's shells. It is not known whether the living female would be able to compensate for this damage.

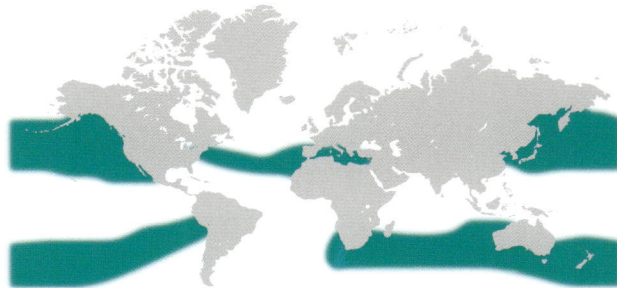

Football Octopus

Tough-bodied mama

SCIENTIFIC NAME	*Ocythoe tuberculata*
FAMILY	Ocythoidae
MANTLE LENGTH	14 in (35 cm)
TOTAL LENGTH	31½ in (80 cm)
NOTABLE ANATOMY	Gas-filled swim bladder
MEMORABLE BEHAVIOR	Internal egg brooding

Like many other pelagic octopuses, including the argonaut, Football Octopuses display extreme sexual dimorphism with relatively large females and miniature males. Cephalopods are not the only pelagic or deep-sea animals to have evolved this trait. A classic example is the anglerfish. In fact, the anglerfish is a helpful model for understanding what's advantageous about this kind of sexual dimorphism.

Animals in the open ocean face a real challenge when it comes to finding a mate. They have no specific location for a spawning ground, no dens in which to seek each other out. So, when they do find a mate, they want to go all in, because the odds of ever finding another are low. Females invest their energy in being large enough to make lots of eggs, and in having a good place to keep them safe. Meanwhile, males put their energy into making lots of sperm and giving it a good shot of getting to a female, neither of which requires them to be very large, so they don't bother investing in a big body.

In a subset of anglerfish species, when a male finds a female, he attaches and fuses with her body, giving up his independence along with most of his organs to become nothing but a sack of sperm for fertilizing her eggs. With the anglerfish for context, the mating activity of pelagic octopuses is not that unusual. Males don't attach themselves to females, but they grow a relatively huge detachable hectocotylus (reproductive arm) that they give to the female. A mature female anglerfish may

have several males hanging off her body, and a mature female pelagic octopus may have several hectocotyli in her mantle.

The female Football Octopus then proceeds to do something that is truly out of character for an octopus: fertilize and brood her eggs inside her body. This species has elongated labyrinthine oviducts that serve as a womb or brood chamber, where the eggs develop all the way to hatching. The mother then proceeds to birth live young.

The female Football Octopus is also completely unique in its possession of a gas-filled bladder, such as fish commonly have to control their buoyancy. Further mysteries: female Football Octopuses have large openings on the underside of their head, into a water-filled cavity, and the females' mantles have a net of cartilage, producing a bumpy appearance and giving them the shape and texture of a football. The purposes of these curious traits remain unknown.

Female

Male

Sexual dimorphism
Male and female Football Octopuses almost seem like different species— the former small, with one huge arm, the latter large, with mantle cartilage and a head cavity.

→   Adult female Football Octopuses are very rarely photographed, as they live at greater depths than their young offspring. The mantle's cartilagenous structure is visible here.

THYSANOTEUTHIS RHOMBUS

Diamondback Squid

Monogamous mate

SCIENTIFIC NAME	*Thysanoteuthis rhombus*
FAMILY	Thysanoteuthidae
MANTLE LENGTH	3¼ ft (1 m)
TOTAL LENGTH	6½ ft (2 m)
NOTABLE ANATOMY	Diamond-shaped fins
MEMORABLE BEHAVIOR	Floating egg masses

The shape of an adult *Thysanoteuthis rhombus* is so distinctive that this species is unlikely to ever be mistaken for another. The same is true of their egg masses. While the gelatinous subsurface egg masses of flying squid are difficult to identify to species, *T. rhombus* is the only squid known to lay an egg mass that floats at or near the surface of the water.

The egg mass's shape has been described as a sausage. Unlike flying squid egg masses, which have eggs studded throughout, or loliginid egg capsules, which are little more than eggs all stuck together, Diamondback Squid eggs are wrapped in helices around the outside of the mass. The egg mass thus resembles a giant strand of DNA, vivid pink in color thanks to the reddish chromatophores of the embryos in the transparent eggs.

This isn't even the only unusual aspect of their reproduction—Diamondback Squid are also the only cephalopod believed to form monogamous pairs that swim and migrate together. Scientists formed this theory in 1998 based on collecting multiple male–female pairs and it has not been disputed since, though it also has yet to be confirmed. Genetic analysis of the 8–12 egg masses laid by a female during her lifetime could be compared to the genetics of the male in the pair to confirm singular paternity. (A 2020 study on the genetics of the Firefly Squid, *Watasenia scintillans*, revealed it to be another contender for a monogamous cephalopod species.)

Diamondback Squid hatch from their eggs as paralarvae that have very tiny fins, which is typical of oceanic squid paralarvae. Their fins are like a miniature toupee on the tip of the mantle. As they grow, the fins expand and change shape, extending further down the mantle in juveniles. As adults, the fins that extend the entire length of the mantle function more like the fins of a manta ray than like the marginal fins of a reef squid or cuttlefish. Diamondback Squid are fairly slow swimmers, more inclined to move by flapping their fins than by the power of jet propulsion.

They are fished around Japan, and live throughout the world's tropical oceans, although they are observed rarely. It may be this scarcity of individuals dispersed over a large area that has led to the monogamous mating system.

→   Although named for the distinctive shape of their fins, Diamondback Squid have numerous unusual features, from their near-surface egg masses to their presumed monogamy.

Purpleback Flying Squid

Glowing ocean-goer

SCIENTIFIC NAME	*Sthenoteuthis oualaniensis*
FAMILY	Ommastrephidae
MANTLE LENGTH	4–20 in (10–50 cm)
TOTAL LENGTH	6½ ft (2 m)
NOTABLE ANATOMY	Large dorsal photophore
MEMORABLE BEHAVIOR	Flying out of the water

Purpleback Flying Squid are found throughout the tropical and temperate Pacific and Indian Oceans. Like other flying squids (Ommastrephidae), they have the ability to jet out of the water and fly for a distance over the surface. For "wings," flying squids use their wide fins, and they also spread their arms and arm webs to increase surface area. Videos of squid flight have revealed that these animals continue to propel themselves forward by squirting water out of the funnel even in the air, which places the behavior into the category of true propelled flight, rather than passive gliding.

Squid flight is probably advantageous for the same reason as the flight of flying fish: a way to escape predators. Flying Squid are themselves hungry consumers of crustaceans, fish, and other squid, but they are also eagerly hunted by many larger consumers, from dolphins and other toothed whales to large fish such as sharks and tuna.

Purpleback Flying Squid come in several distinct "morphs" that may actually constitute cryptic species. They are known for their large dorsal photophore (dorsal means

"on the back"), but the smallest form, found only near the equator, lacks this photophore. These individuals are mature adults at only 4 in (10 cm) mantle length. The "typical" form found throughout the species' range matures at around 6–8 in (15–20 cm) mantle length and has the photophore, as does the "giant" form found only in the Red and Arabian Seas, reaching up to 20 in (50 cm) mantle length.

Like the closely related Humboldt Squid (*Dosidicus gigas*), which overlaps with Purpleback Flying Squid in part of their ranges, the wide variation in size at maturity could be caused by variable environmental influences—fluctuations in temperature and food availability. However, genetic analysis has shown significant divergence between groups of Purpleback Flying Squid.

A giant Purpleback Flying Squid gained notoriety in 2021 for "photobombing" a research expedition in the Red Sea. The crew had already been surprised by the unexpected discovery of a large shipwreck; they were subsequently shocked by the sudden appearance of an enormous squid exploring the wreck. Although Purpleback Flying Squid are found around the world and regularly fished, often for use as bait, we know extraordinarily little about them. Thus, the encounter with a large individual hanging around on a shipwreck provided new and intriguing information—flying squid were thought to be truly open ocean animals, but apparently they can also use underwater structures as habitat.

→   Using models of the Purpleback Flying Squid to study the process of launching from water to air during squid flight, researchers have learned that the lower the launch angle, the faster the flight.

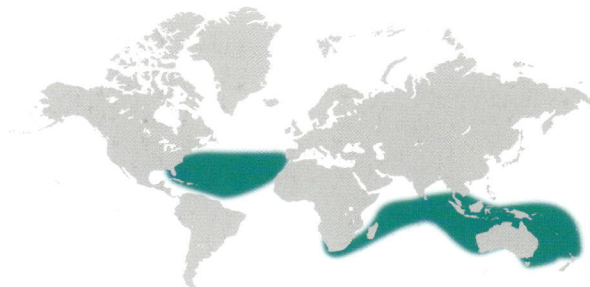

Ram's Horn Squid

Secret coil

SCIENTIFIC NAME	*Spirula spirula*
FAMILY	Spirulidae
MANTLE LENGTH	1¾ in (4.5 cm)
TOTAL LENGTH	2¾ in (7 cm)
NOTABLE ANATOMY	Coiled internal shell
MEMORABLE BEHAVIOR	Head retraction

The Ram's Horn Squid, with its eight short arms and two long tentacles, is related to squid and cuttlefish. But all three groups had already diverged from each other well before the end of the Cretaceous, 66 million years ago. Fossils of early spirulids have a straight internal shell that is only coiled in the early part of life. Modern spirulids, which continue to grow coiled shells into adulthood, are thought to have evolved through neoteny—the evolutionary process of retaining youthful features for more and more of the life cycle, a story we'll see again with Dana's Chiroteuthid Squid, *Planctoteuthis danae* (pages 208–209).

In their open ocean habitat, ancestral spirulids would have been preyed upon by early whales and dolphins. When echolocation came on the scene, over 30 million years ago, cetaceans acquired the ability to hunt for food without having to see it. Visual camouflage and ink smokescreens that cephalopods had evolved for defense against fish predators could not hide them from echolocation. And their hard, gas-filled shell was a resonant target that sent back a strong auditory signal. It's possible that the once-diverse group of spirulids fell prey to cetacean hunters. Those that survived may have done so by shrinking both their internal shells to small tight coils, and their entire bodies to a size no longer worth targeting by a large cetacean.

Preyed upon now by fish and seabirds, the Ram's Horn Squid hides with classic visual defensive tools: color-changing skin and a photophore on the tip of its mantle that that the squid orients downward, canceling out its shadow against the light background of the surface (see page 163).

Ram's Horn Squid also exhibit a hiding behavior of retracting the head into the mantle, like a snail or a turtle pulling into its shell. Flaps on the mantle can enclose the head fully inside. Since the animal's body is not a hard protective shell, the advantage of this behavior is not immediately clear—perhaps it simply presents a smaller target.

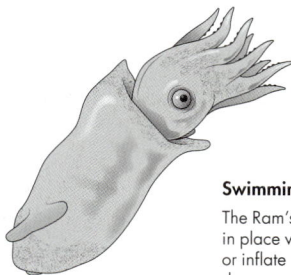

Swimming *Spirula*
The Ram's Horn Squid can hover in place with constant fin movement, or inflate its mantle for a quick downward escape jet.

→ Ram's Horn Squid are most widely known from their shells, which often wash up on beaches. The animals are sometimes captured in nets and photographed dead, while live animals in their natural habitat were only recorded (in low resolution) for the first time in 2020.

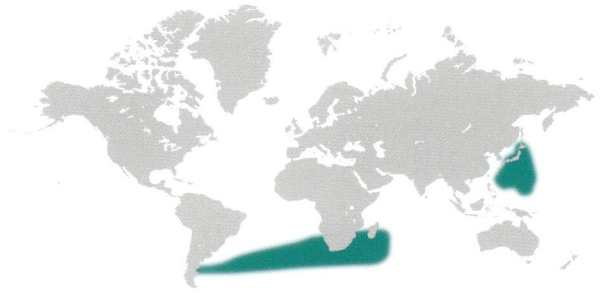

Large Glass Squid

Transparent giant

SCIENTIFIC NAME	*Megalocranchia sp.*
FAMILY	Cranchiidae
MANTLE LENGTH	6 ft (1.8 m)
TOTAL LENGTH	9 ft (2.7 m)
NOTABLE ANATOMY	Gelatinous skin layer
MEMORABLE BEHAVIOR	Possible reverse migration

A typical glass squid in many ways, *Megalocranchia* lives near the surface as a paralarva and moves deeper as it ages. Mature adults, however, have been caught near the surface, leading researchers to guess that they undergo a second, reverse migration to spawn in shallow water. Young individuals are transparent and have a curiously thick skin containing a gelatinous layer, with chromatophores both above and below this layer. Scientists complain that this slippery feature makes it difficult to pick up the animals when they are caught in a net, which suggests that it may also help young *Megalocranchia* evade predators.

The exact number of species in this genus is still being worked out, and it's not always clear which species any given specimen belongs to.

Megalocranchia has photophores under its eyes, a typical glass squid trait to counteract the shadows cast by these solid body parts. The tips of *Megalocranchia* arms are also adorned with photophores—but in females only. We've seen sex-specific arm tip decorations before, as in the case of the modified suckers of Star-sucker Pygmy Octopuses (*Octopus wolfi*, pages 132–133), and their purpose is generally assumed to be related to courtship or mating, but these behaviors have never been observed in *Megalocranchia*. They are also the only glass squid known to have photophores on their digestive glands.

As one of the largest squid species, *Megalocranchia* has been implicated in at least one report of squid sucker scars left on a large shark. This particular whitetip shark was observed in Hawaiian waters, and the rows of sucker marks visible on its skin had to belong to one of only three huge squid in the area: the Giant Squid (*Architeuthis dux*), the Diamondback Squid (*Thysanoteuthis rhombus*), or the Large Glass Squid. In general, the shapes of squid suckers are not distinctive enough for the marks they leave behind to be identified to species. Sucker marks observed on animals such as sharks or whales are always evidence of those animals trying to eat squid, never the other way around. While squid may well put up a good fight, there's no evidence that any size of squid could eat something as large as a whale or a whitetip shark. (Although there is at least one account of an octopus eating much smaller species of sharks in an aquarium.)

Glass squid giant

Most glass squid are relatively small. *Megalocranchia*, which can grow longer than an adult human, is a startling exception.

→ The genus *Megalocranchia* is currently thought (as of 2022) to contain three species, *M. maxima*, which is subtropical and temperate; *M. oceanica*, which is tropical; and *M. fischeri*, shown here as a paralarva from Hawaiian waters. The transparent gelatinous skin layer can be seen surrounding the mantle.

Strawberry or Cock-eyed Squid

Spangled spectacle

SCIENTIFIC NAME	*Histioteuthis heteropsis*
FAMILY	Histioteuthidae
MANTLE LENGTH	5 in (13 cm)
TOTAL LENGTH	Estimated 10¼ in (26 cm)
NOTABLE ANATOMY	Red skin dotted with photophores and unequal eyes
MEMORABLE BEHAVIOR	Blinking

The genus *Histioteuthis* contains numerous species of cock-eyed squid, all possessing one large eye and one small. These eyes are specialized to perceive different sources of light: the large one peers up to catch all possible traces of sunlight from the surface, while the small one looks down to perceive flashes of bioluminescence from fellow squid and potential prey items. Paralarval cock-eyed squid are born with symmetrical eyes, but as the animal grows, so does the left eye, much faster than the right. In some but not all individuals, the large left eye has additional yellow pigmentation that may serve as a screen to block out counter-illumination by their prey. And so the oceanic arms race continues.

The name Strawberry Squid is specific to *Histioteuthis heteropsis* and derives from the animal's typical red coloration and seed-like pattern of photophores. Of course, it would not look red at the depths it normally lives, because red light is filtered out of seawater so quickly. It only looks red when brought to the surface, or illuminated by the lights of an ROV (remotely operated vehicle). Although most open ocean squid have photophores, the densely dotted photophores

of Strawberry Squid are unusual. They may be used to communicate with other members of the species, as well as for camouflage and perhaps to attract or distract prey.

Strawberry Squid live on the eastern side of the Pacific Ocean, both north and south of the equator but not at the equator itself, a distribution known as antitropical. They seem to be adapted to the two cold-water currents that sweep along the temperate coasts of North and South America, respectively: the California Current and the Humboldt Current. Strawberry Squid are diel migrators that are abundant enough to be often observed by ROVs and to be often consumed by local vertebrate predators such as hammerhead sharks and elephant seals. However, they remain mysterious enough that we still don't know much about key aspects of their biology, like reproduction.

The Strawberry Squid is somewhat famous among oceanic squid for having been recorded blinking its eye. A brief video taken by a submersible in 2013 is one of the only visual records of the oegopsid eyelid, confirming that these squid without corneas have flaps of skin to protect their eyes.

→ The Strawberry Squid's large left eye is clearly visible in this photo, while its smaller right eye is out of view on the other side of the animal. The smaller red dots on its skin are chromatophores and the slightly larger darker spots are photophores, not currently illuminated.

MIDWATER

The twilight zone

Below the sunlight zone in the open ocean is the twilight zone, where some light from the surface still reaches, but not enough for photosynthesis to take place. Past the twilight zone is the fully dark midnight zone. The depth to which light penetrates depends on many factors, so the names and depths of these zones are not set in stone but rather written in water.

THE OCEAN AS A LAYER CAKE

Another set of names for dividing up the open ocean are epipelagic, mesopelagic, and bathypelagic. "Epi" refers to surface (as in epidermis, the skin at the surface of your body), "meso" means middle, and "bathy" means deep. In the previous chapter, we considered primarily the epipelagic zone. Here, we'll explore the mesopelagic, and in Deep Sea (pages 212–247), we'll venture into the bathypelagic and below.

The general rule of thumb is that enough light for photosynthesis is gone by 660 ft (200 m), and we humans typically can't perceive any sunlight below that depth. But photons from the sun can penetrate as far as 3,300 ft (1,000 m) below the surface, if the water is clear, and some midwater animals have eyes sensitive enough to pick up these tidbits of light. Thus, from 660 to 3,300 ft (200–1,000 m) can be referred to as the twilight or mesopelagic zone. Deeper than 3,300 ft (1,000 m) is the midnight or bathypelagic zone, where sunlight is truly gone. Many of the organisms we'll meet in this chapter move between zones, as our linguistic distinctions don't constitute biological boundaries. The deep scattering layer that engages in diel migration (see pages 150–153) routinely moves from epipelagic to mesopelagic layers, for example. We'll use the most general term, midwater, to cover all the space that is neither the surface nor the seafloor depths.

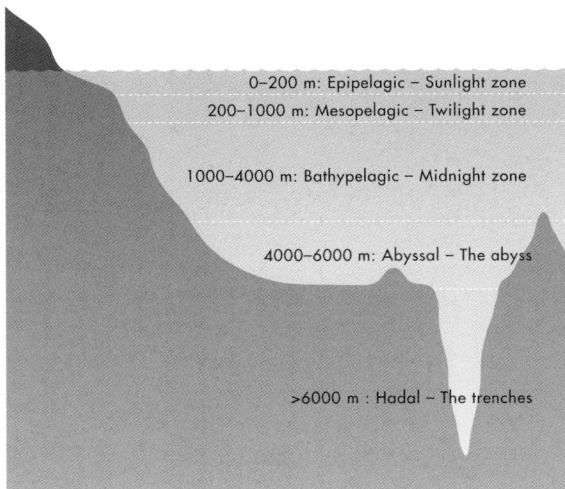

0–200 m: Epipelagic – Sunlight zone
200–1000 m: Mesopelagic – Twilight zone

1000–4000 m: Bathypelagic – Midnight zone

4000–6000 m: Abyssal – The abyss

>6000 m : Hadal – The trenches

Ocean zones

The attentuation of light in the ocean varies from location to location and season to season, depending on the amount of dissolved and suspended material in the water. However, scientists have settled on constant depths to represent the boundaries between sunlight, twilight, and midnight zones.

→ This see-through Glass Octopus shows the typical organization of a cephalopod brain: one central mass and a large optic lobe between each eye and that mass.

MIDWATER VISION

Eyes that have evolved to pick up tiny pinpricks of light from the surface, as well as the brief flashes of bioluminescence in this region, are often bizarre from our light-drenched land-dwelling perspective. In the previous chapter, we met the Strawberry (Cock-eyed) Squid (*Histioteuthis heteropsis*), with its single enormous eye. Bigger eyes can usually capture more light, and the midwater is home to some truly bug-eyed creatures.

Eyestalks are another trait midwater cephalopods have in common with insects. The Telescope Octopus (*Amphitretus pelagicus*) is named for its cylindrical peepers. The Sandal-eyed Squid (*Sandalops melancholicus*), meanwhile, begins life as an epipelagic paralarva with eyes shaped like the soles of shoes. When it grows and sinks to beome a mesopelagic juvenile, its eyes become tubular, pointing upward to get every bit of light. As an adult living deeper still, its eyes widen into hemispheres, no longer straining for sunlight but merely perceiving bioluminescence. Firefly squid may have the most astonishing midwater vision of all: they're the only cephalopods known that can possibly see colors. Their multiple kinds of photoreceptor cells are similar to our cones but tuned to different wavelengths, all of which we would call blue. The fine-grained ability to distinguish between blues may serve them well at a depth where few or no other colors are present.

↑ This abundance of
myctophids, or lanternfish,
would be a Humboldt
Squid's buffet. These fish
are fundamental to the
midwater ecosystem,
feeding off zooplankton
and providing meals for
larger predators.

A PLETHORA OF FISH

The midwater, considered broadly, is the largest habitat on the planet, and it contains the greatest abundance of animals. We can break this down by considering representatives of three large marine groups: vertebrates, crustaceans, and, of course, cephalopods.

Among vertebrates, the lanternfish (Myctophidae) and bristlemouths (Gonostomatidae) dominate the midwater. Scientists estimate that lanternfish make up 65 percent of all deep-sea fish biomass, while a single genus of bristlemouths, *Cyclothone*, is the most abundant vertebrate genus on the planet. The ocean is estimated to contain one quadrillion individuals of *Cyclothone*.

These fish are small, but not too small for humans to catch and eat. Both groups are bioluminescent, with the lanternfish having been named for it, but the bristlemouths were named instead for their enormous prickly maws. Their bioluminescence may help them with counter-illumination, and is also likely a mechanism of communication with one another.

Both groups of fish migrate vertically and constitute major components of the deep scattering layer. Small as they are, accomplishing their daily migrations by muscle power alone would be a challenge. Luckily, these fish have a buoyancy device that helps to lift and lower them: a swim bladder. In the previous chapter, we heard about the Football Octopus, *Ocythoe tuberculata* (pages 166–167), the only cephalopod species with a swim bladder. That animal has been so little studied that we don't really know how it uses this organ, but scientists were able to recognize it because of the extensive research on fish swim bladders.

↖ This bristlemouth fish displays the trait for which its group is named: an extremely toothy mouth. Although it appears fearsome, it is only a couple of inches long.

↗ This midwater shrimp broods eggs under her abdomen while using her long antennae to hunt for food.

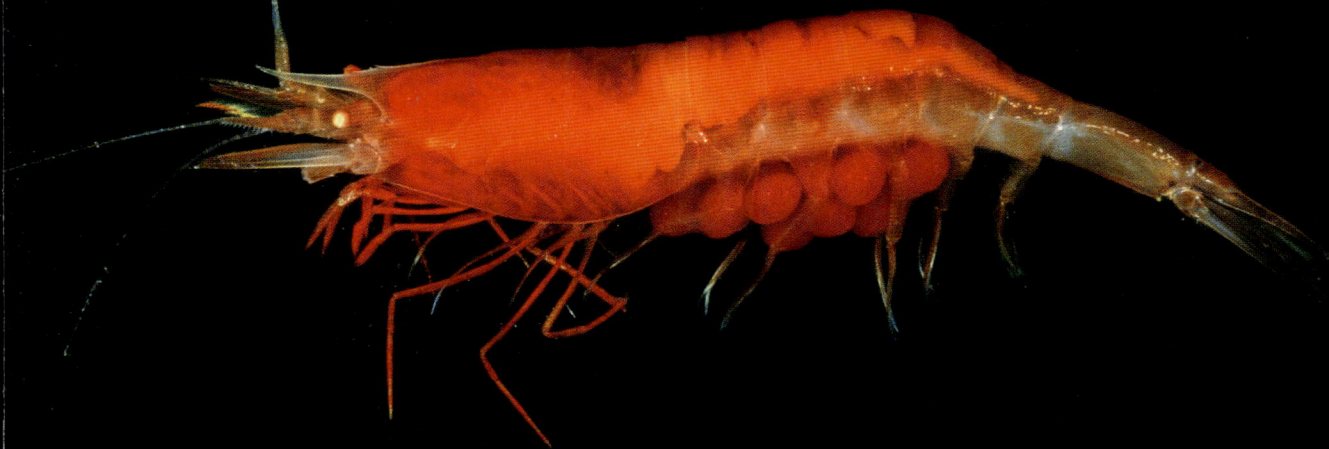

When they are ready to rise from the mesopelagic zone, fish with swim bladders can make dissolved gas diffuse out of their blood and into their bladder (rather like the technique of a nautilus filling its chambers), which then raises them like a buoyant balloon. However, as they go up and experience reduced water pressure, the gas expands further, and they must exert active control to prevent the bladder from bursting. Some lanternfish shift from a gas-filled swim bladder to a fat-filled swim bladder as they grow, which provides consistent neutral buoyancy so the fish don't have to worry about exploding. Midwater fish consume mostly crustaceans, which brings us to the vast abundance of this group.

A CAST OF CRUSTACEANS

Midwater crustaceans include many different groups that we tend to lump together as "shrimp." They are larger than the grazing copepods we met in the previous chapter and exist at least one step up the food chain; in fact, many prey on copepods. Like midwater fish, midwater crustaceans have bioluminescence for counter-illumination, and take part in the grand diel migration.

Midwater shrimp include krill, which we'll hear about more in Antarctica and the Arctic (pages 248–277), but it's worth noting now that they are the main source of nutrition for baleen whales—the group of cetaceans that grow to massive sizes by filtering enormous quantities of tiny plankton out of the water. Because they don't have sharp teeth and don't behave in typically "predatory" ways (that is, they don't hunt and shred seals like killer whales do), we tend to think of baleen whales as grazers, like giant copepods. We have to remind ourselves that they are actually predators, too. They may accidentally consume algae from time to time, but the bulk of their diet consists of tiny but living animals, trapped in the mesh of their mouth.

Oxygen minimum zones

We've already seen how oxygen can be depleted at depth as a result of algal blooms sinking and decomposing. This process happens constantly, even in the absence of extreme blooms. Wherever there's life near the surface, wherever there are plankton and predators growing and living and dying, there will be a steady rain of sinking debris—feces, dead bodies, shed exoskeletons. This marine snow supports midwater and deep-water organisms, from microscopic decomposers to relatively large scavengers such as the Vampire Squid (*Vampyroteuthis infernalis*).

ORIGINS OF LOW OXYGEN

Oxygen minimum zones (OMZs) are created in locations that have plenty of organisms using up the available oxygen, no photosynthesizers to add more oxygen, and limited mixing with other layers of water. You might initially think that the deeper you go, the less oxygen there would be, until the lowest oxygen levels of all would be at the greatest depths. However, the consumption of marine snow in the midwater is so efficient that not much matter is left to fall below about 3,300 ft (1,000 m). Thus, the decomposition process that uses up so much oxygen slows down at greater depths. Furthermore, because water in the ocean is constantly being moved around on a huge scale by a process known as the ocean conveyor belt, the deep sea is fairly well oxygenated.

This happens because of a dynamic at the poles whereby cold, oxygen-rich surface water sinks to the seafloor. This dense water spills over into the basins of the Pacific and Atlantic and flows through all the deep sea (see pages 212–247). Thus, below about 3,300 ft (1,000 m), oxygen levels rise again, leaving the ocean's lowest oxygen concentrations in midwater. These OMZs are not present everywhere throughout the world's seas, but are found primarily in parts of the world with more photosynthesis near the surface, such as the eastern Pacific and Indian Oceans.

↗ Closely related species can evolve adaptations to different oxygen levels. Polar krill in highly oxygenated waters, like this one, keep their oxygen consumption constant, while tropical krill that encounter oxygen minimum zones modify their consumption based on what's available.

← The Vampire Squid's ancestors likely lived in shallow, well-oxygenated waters, like many octopuses do today. But modern Vampire Squid have specialized adaptations to deep water with very little oxygen.

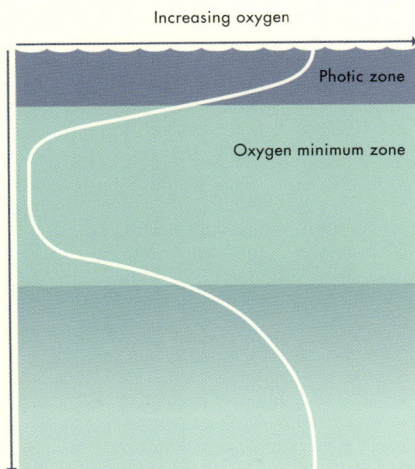

HOW DO ANIMALS COPE?

Many bacteria don't require oxygen at all, or even find it toxic, so an abundance of anaerobic (no-oxygen) life thrives in the OMZ on the microscopic scale. But what about animals, a group that evolved to depend on oxygen? Most of the sea's large and active swimmers, such as swordfish and sharks, simply avoid OMZs.

However, a variety of adaptations are possible to cope with low oxygen. One is metabolic suppression, or simply reducing the body's need for oxygen, which can be done by reducing activity (like a hibernating squirrel). Another option is to switch to anaerobic metabolism, or getting energy without oxygen. Animals can typically use anaerobic metabolism in the short term, as humans do during exercise, but can't sustain it for very long. OMZ-adapted animals can sometimes employ it for much longer than other species.

A third adaptation is to evolve anatomy and physiology for extracting the small amount of available oxygen with maximum efficiency. Many OMZ fish and crustaceans have evolved respiratory and circulatory systems to do exactly this. They breathe rapidly, increasing the amount of water passed over their gills; they also pump their blood quickly, speeding up the transfer of oxygen from the water to the tissues; and their gills have an unusually high surface area and very thin membranes across which the oxygen can diffuse.

The OMZ

Life near the surface produces organic material that sinks, like feces and dead bodies. Decomposition of this material at depth sucks up oxygen, producing a layer called the oxygen minimum zone.

Increasing oxygen

Photic zone

Oxygen minimum zone

Increasing depth

CEPHALOPODS AND LOW OXYGEN

Cephalopods, as we've seen, have high oxygen needs due to their typically active lifestyle and their overall physiology. Squid in particular need more oxygen than fishes and mammals in any given situation. Yet numerous cephalopods have evolved adaptations to thrive in the OMZ.

Vampire Squid consume marine snow and live out their lives in low-oxygen waters. These two features are related—the passive collection of falling particles doesn't require nearly as much movement as active hunting, so Vampire Squid can afford to suppress their

metabolism. In fact, they have the lowest metabolic rate (relative to their mass) ever measured for a cephalopod.

And they're not the only cephalopods to engage in metabolic suppression in OMZs. Amazingly, the extremely active Humboldt Squid (*Dosidicus gigas*) has also been found to thrive in OMZs. This species

↑ Humboldt Squid are large, active predators, growing up to six feet and hunting a wide variety of crustaceans, fish, and other cephalopods. Their ability to cope with low oxygen came as a surprise.

migrates nightly into well-oxygenated surface waters, then spends its days at around 985 ft (300 m) depth, in the middle of the eastern Pacific OMZ. During daytime hours, these squid suppress their metabolism and reduce their activity levels.

At low-oxygen depths, Humboldt Squid cannot hunt actively as they do near the surface, but they benefit by conserving energy and avoiding other large predators, such as tuna and sharks. It's possible they are still capable of grabbing a meal if convenient, as remotely operated vehicles (ROVs) have recorded Humboldt Squid hunting lanternfish while in the OMZ. However, these observations may not represent typical behavior, and were possibly triggered by ROV lights (see pages 190–191). As a rare large predator that can take advantage of the OMZ, the expansion of this habitat due to anthropogenic influences may allow Humboldt Squid to expand their range as well.

CHANGES IN GLOBAL OMZS

Over the long history of Earth's oceans, the amount and extent of seawater oxygenation has changed many times, often driving extinction or adaptation. We know that overall oxygenation in the ocean is decreasing as the water warms due to climate change, because warm water cannot hold as much dissolved oxygen. This reduction in baseline available oxygen will cause OMZs to grow.

Evidence suggests that two changes are occurring in global OMZs: expansion, which means growth of the geographic area they cover, and shoaling, which means they're getting shallower. Off the Pacific coast of North America, shoaling of the OMZ has been measured as great as 295 ft (90 m) since 1984. The animals that have adapted to these habitats find themselves with an increased available range, while those that cannot tolerate low oxygen are compressed into a smaller range.

This isn't great news for biodiversity, since a lot of the richest coastal ecosystems are losing ground against OMZs. It's also not great news for fisheries. Although some midwater OMZ animals are targeted by humans (Humboldt Squid support one of the largest squid fisheries in the world), OMZ animals tend to be less muscular and more gelatinous, and therefore less nutritious and appealing to catch.

OMZ locations

The global distribution of oxygen minimun zones (OMZs) is determined by patterns of water movement and biological activity. They predominate in tropical and temperate oceans at the western edges of continents.

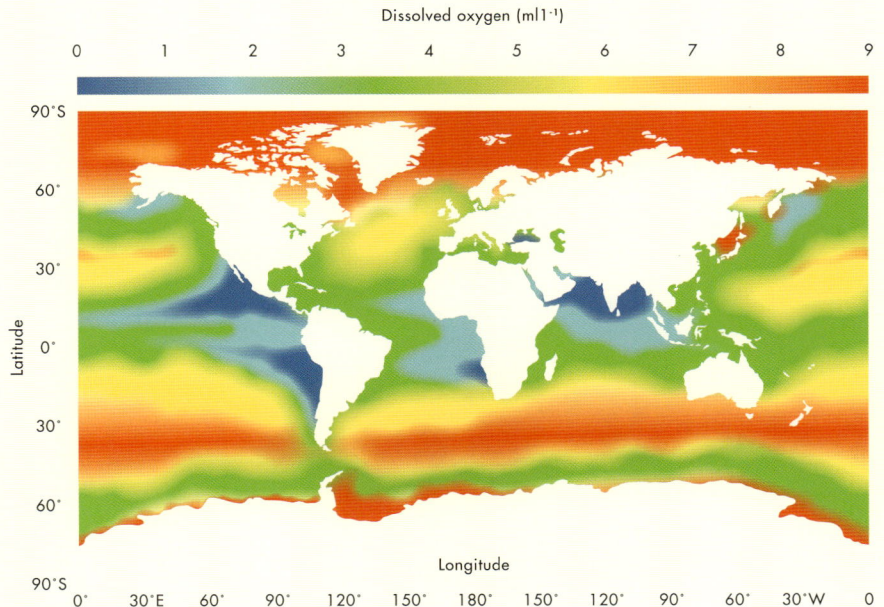

Dissolved oxygen (ml l⁻¹)

Submersibles

Although many midwater organisms spend parts of their lives near the surface, others remain all their lives at greater depths. Historically, people could only study these animals when they were caught by fishing gear, limiting what could be learned. Behaviors weren't observable, and even anatomy was difficult to comprehend—soft, gelatinous bodies lose their shape and structure when brought to the surface. Submersibles presented an enormous advance in our ability to understand this part of the world.

TAXONOMY OF SUBMERSIBLES

There are three types of submersibles: human-occupied vehicles (HOVs); remotely operated vehicles (ROVs), which are controlled through a physical tether to a ship; and autonomous underwater vehicles (AUVs), which have no humans or surface attachment, and are operated by software.

ROVs are used more frequently than HOVs, since they can be smaller, stay underwater longer, and are otherwise unconstrained by the complications of keeping air-breathing mammals alive. ROVs have facilitated many discoveries of new species and behaviors—such as the detritus feeding of Vampire Squid.

VAMPIRE SQUID ROV STUDY

This squid received its common name as well as its scientific name (*Vampyroteuthis infernalis*, literally "vampire squid from hell") in 1903, when its feeding habits were completely unknown. It had webbed arms like a cloak and eyes that sometimes looked red, and that was enough to merit a monstrous name.

← Whiplash squid are named for both the length and the shape of their tentacles. Instead of a wide tentacular club, the tips of their tentacles are slender and covered with tiny suckers.

ALVIN

....

One of the most famous HOVs is *Alvin*, of the Woods Hole Oceanographic Institution in Massachusetts, in which scientists have been diving to the midwater and deep sea since 1964, and which brought the first human observers to the wreck of the *Titanic* in 1986. Development of HOVs to explore the deep sea proceeded in tandem with, although with less media coverage and excitement than, the development of spacecraft to explore outer space.

Then, between 1992 and 2012, ROVs accumulated enough video footage of Vampire Squid for scientists at the Monterey Bay Aquarium Research Institute (MBARI) to figure out how these cephalopods eat. They saw that two slender filaments, which are highly modified arms, stretch out one at a time to collect particles, then bring back to the mouth. Fully extended, the filaments can be eight times as long as the Vampire Squid's body.

Ventana and *Doc Ricketts*, two ROVs operated by MBARI, also collected live Vampire Squid with suction samplers that use negative pressure to pull an animal into a plastic water-filled container. This prevents abrasion and minimizes collection damage.

Suction sampling is so effective that blobs of food and mucus about to be eaten, or regurgitated by the animal, could also be collected. This is how scientists found that the Vampire Squid diet includes everything from mucus discarded by other organisms to pieces of jellyfish and crustaceans to eggs and fecal pellets of copepods.

SUBMERSIBLES IN THEIR ENVIRONMENT

Submersibles allow observation of animals in their natural habitat, but they are capable of affecting that habitat and modifying behavior. If the bright lights sported by most submersibles illuminate prey, then visual predators such as squid may take advantage of the situation to hunt, even if they might not normally do so.

The ROV itself can attract the attention of deep-sea organisms, especially cephalopods, with their ability to investigate their surroundings. The ROV *Deep Discoverer* had one such encounter in 2020 in northwest Hawaiian waters. A whiplash squid (Mastigoteuthidae) gripped the ROV for a few minutes and rode along, in addition to swimming in front of the cameras long enough for the scientists to get a good view.

ROV exploration has brought many other cephalopod observations to light. The Schmidt Ocean Institute's ROV *SuBastian* recorded the first video of Ram's Horn Squid (*Spirula spirula*) in their environment, confirming their head-up, mantle-down orientation to use their light organ for counter-illumination.

Anatomical extremes

Sometimes we think of the deep sea as the home of the ocean's most fabulous weirdos (and there are plenty of deep-sea oddballs, as we'll see in the next chapter), but the deep sea has a seafloor, which makes for what we land-dwellers think of as a "normal" habitat. Octopuses that crawl along this seafloor live at incredible depth, pressure, and darkness—but at the same time, their lifestyle isn't all that different from an octopus crawling along the sand near a coral reef. Their anatomy reflects this similarity. By contrast, the middle of the ocean, with no surface and no floor and no reef-like or algal structures, constitutes an environment that is probably the most alien to us humans of any environment on the planet. The animals here get pretty alien-looking, too.

THE BABY THAT LOOKS LIKE A TOXIC JELLY

Let's start with the doratopsis paralarva, the young stage of a group of squid known as Chiroteuthidae. The full extent of their weirdness wasn't even known until ROVs began exploring the midwater, because their most distinctive feature is too fragile to be collected by a net. Doratopsis paralarvae have a "tail" that is an enormous extension of the tip of the mantle, on the end that has the fins, opposite the end with the head and arms hanging out. This tail can be several times longer than the rest of the body and is extremely straight and stiff, since its structure is maintained by an extension of the gladius (a squid's internal shell remnant). The gladius itself is fragile and easily broken, however, by any kind of rough contact—which goes a long way toward helping us understand just how infrequent physical contact is in the midwater zone. This is an area where you spend all your life touching nothing except your food and one day, hopefully, a mate.

→   This adult chiroteuthid squid is certainly distinctive: its tentacles and arms are exceptionally long and its head is connected to its mantle by a long neck. However, its babies are even stranger.

↓   These doratopsis paralarvae of chiroteuthid squid demonstrate a bizarre elongated "tail," as well as a very long "neck" between head and mantle.

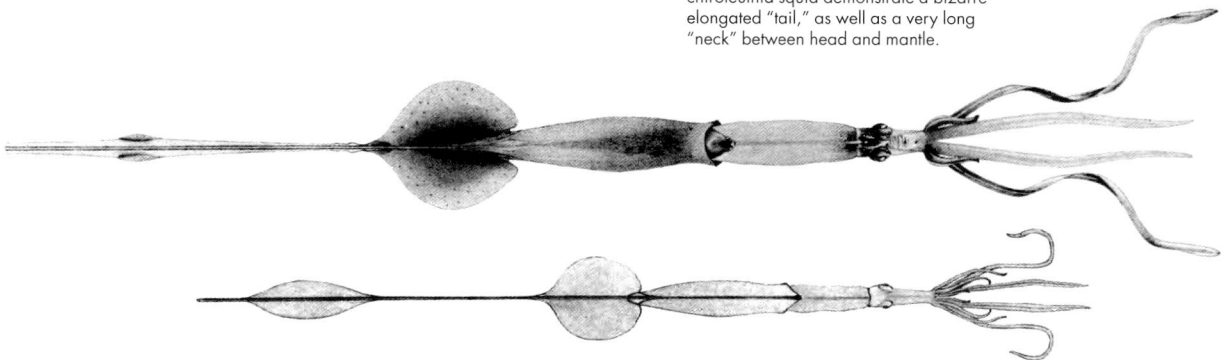

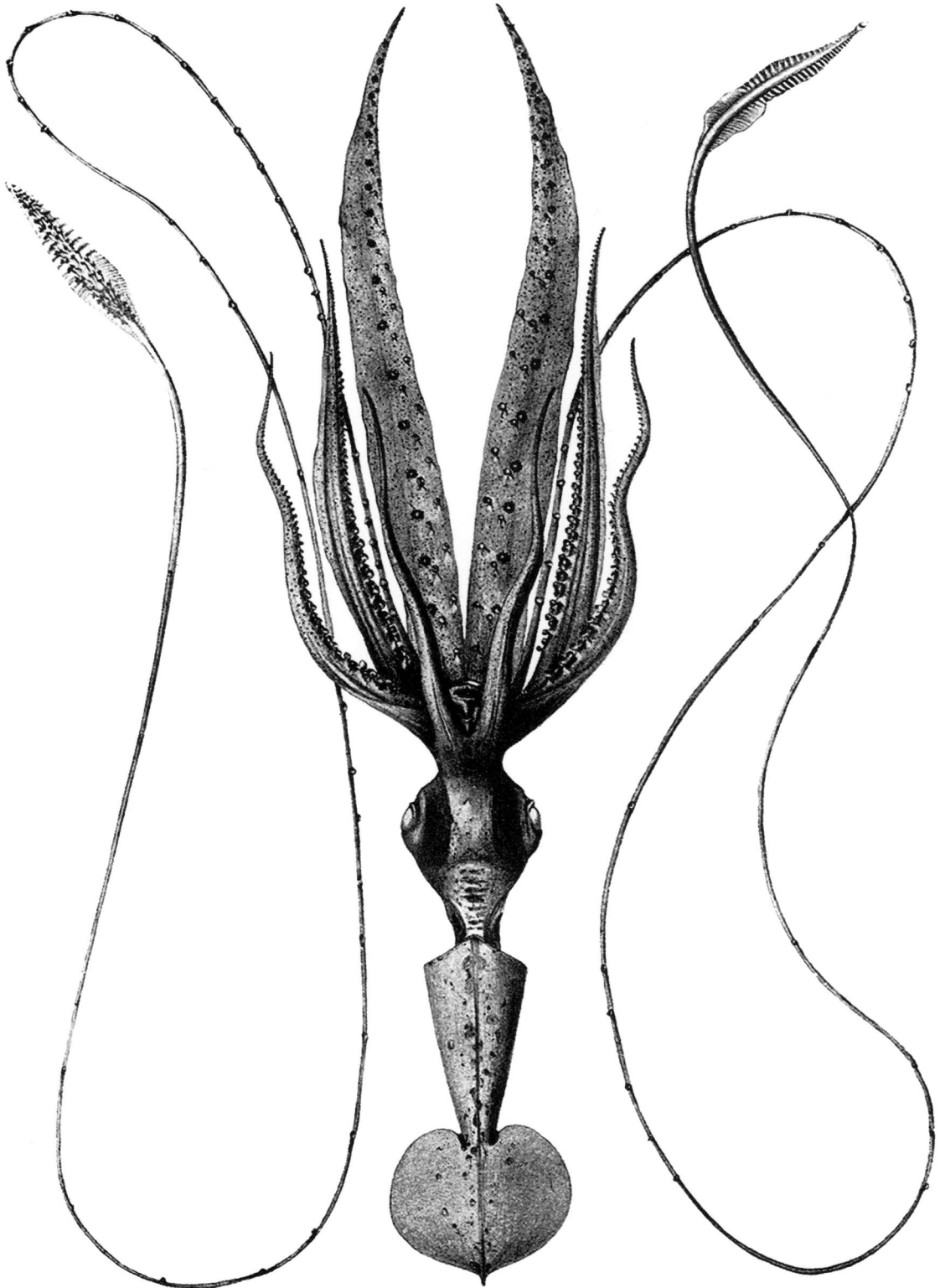

The doratopsis tail is not only a long thin rod. It is fantastically decorated with a variety of ornaments, including what appear to be extra pairs of fins, many small flaps, and even bulbs that look like the gas bladders of kelp, which are filled with a lightweight fluid that provides buoyancy. The skin of the tail is fully equipped with color-changing organs, chromatophores and iridophores, and ROVs have recorded striking color displays.

Whenever an ROV happens across a doratopsis paralarva, unless a cephalopod specialist is present, there is usually considerable confusion over what kind of animal is being observed. It's likely that we humans are not the only animals to react this way! Scientists think that the long doratopsis tail is a form of mimicry, and these paralarval squid have evolved to look like a different and much less tasty animal: a siphonophore.

Siphonophores are cnidarians, in the same group as coral and anemones and jellyfish. They live in colonies, like coral, but unlike coral each colony contains both medusa and polyp stages. Also unlike coral, the members of a siphonophore colony, called zooids, are specialized in their tasks. Some zooids take care of the eating, others handle the protecting, others produce gas for buoyancy, and still others do the reproducing. Like squid, siphonophores are carnivorous; unlike squid, they possess the potent stinging cells typical of cnidarians. This is presumed to be the main benefit to squid paralarvae of mimicking them. Very few animals actually prey on siphonophores, due to the toxic nature of their tentacles. (The Portuguese man o' war, *Physalia physalis*, is a siphonophore known to have killed humans with its sting.)

↖ Unlike jellyfish, which are single individuals, siphonophores are colonies of polyps. They often grow long stems with a variety of polyps for feeding, floating, and reproduction.

← The gills of an adult axolotl evolved through neoteny from the gills of their tadpoles.

Doratopsis paralarvae occupy the same habitat as siphonophores, including both the epipelagic and the upper mesopelagic. As they mature, they lose the tail and sink to greater depths, including both lower mesopelagic and bathypelagic zones—although they never become anything close to benthic.

One genus of chiroteuthid squid, *Planctoteuthis* (pages 208–209), has evolved to retain the doratopsis tail into adulthood. They also don't grow as large as other members of their family. This is thought to be an example of neoteny, an evolutionary process whereby a species keeps larval or juvenile traits that were previously lost during maturity. Neoteny is perhaps most famous from the axolotl salamanders that keep their larval gills as mature adults, but it has been observed in a variety of species, including Ram's Horn Squid (*Spirula spirula*, pages 172–173) and potentially Glass Octopuses (*Vitreledonella richardi*, pages 200–201). The reasons for neotenous development in *Planctoteuthis* remain unknown.

COMB-FINNED SQUID

Like the chiroteuthids, here is a squid with a completely different take on a standard part of the standard squid body. Instead of an elongated tail, *Chtenopteryx* species grow fins with ribs like the teeth of a comb. Also like the chiroteuthids, their anatomy wasn't clear from early damaged specimens that had been pulled up onto boats.

Comb-finned squid have long fins that reach nearly the entire length of their mantle, resembling the fins of cuttlefish and reef squid. Each fin is lined with stiff ribs, making them look rather like the appendages of ray-finned fish. The membranes connecting these ribs were often damaged and torn in historical collections, which is why early drawings of chtenopterygids make the fins look even more like combs, with separate unconnected "teeth." The ribs are not cartilage or bone, shells or shell vestiges, but simply very stiff muscle fibers.

The paralarvae of comb-finned squid do not have a special name, but they do have their own unusual tentacles, like many other midwater cephalopod paralarvae, such as the doratopsis and rhynchoteuthion. The tips of the tentacles of comb-finned squid paralarvae, instead of being flattened into the palm-like club typical of squid, end in a disk dotted with suckers. Paralarval behavior has not been observed, so we don't know what they do with these unusual tentacles. Residents of tropical and subtropical seas around the globe, comb-finned squid in the limited locations where they have been studied appear to engage in vertical migration.

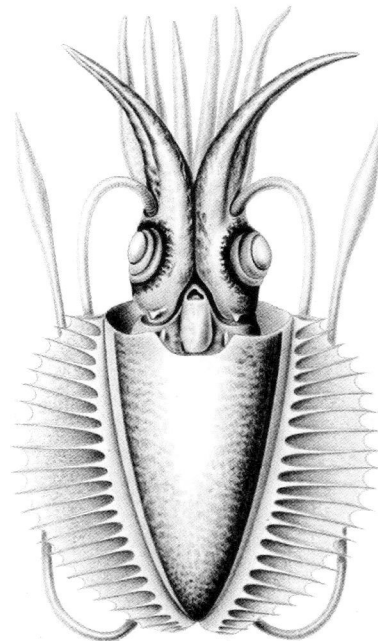

→ Why comb-finned squid evolved such sturdy supports on their fins remains a mystery; luckily for scientists, it makes them easy to identify.

THE GIANT SQUID

What about Giant Squid (*Architeuthis dux*)—do they have strange paralarvae? Amazingly, we have yet to collect or observe paralarvae of this enigmatic species. We may well discover that *A. dux* paralarvae have their own peculiar tentacle adaptation, or other oddities yet to be uncovered. There is only one species of Giant Squid according to current research, although as many as 20 have been named over the years, with overeager researchers often erecting a new species from a single specimen. Careful anatomical and genetic research to date indicates that the variation is all contained within a single species.

While hardly commonplace, adults and juveniles of Giant Squid are not so difficult to find as paralarvae, and indeed Giant Squid are a periodic nuisance in large fishing nets in Japan, where they sometimes become caught and cost fishers significant time and effort to release. It is around the waters of Japan that living Giant Squid have most often been observed, since they seem to live shallowest and closest to shore here.

However, Giant Squid are certainly no naturally nearshore animals. When individuals are seen on or near the beach, or swimming close to a dock, they are disoriented and unwell, having been carried to the unfamiliar location by currents of too-warm water. Their natural habitats are the mesopelagic and bathypelagic depths. We have very little idea of what they do. They've never been successfully filmed by an ROV, but footage of Giant Squid in their natural habitat has been obtained by dropping cameras with bait, so we know they can hunt actively and swim fairly fast. We also know they are incredibly abundant, simply based on the number of Giant Squid beaks found in the stomachs of their main predators, sperm whales. Given this abundance, it's astonishing that we haven't yet spotted their eggs, paralarvae, or large aggregations of adults. Clearly we have a great deal more to learn about this species.

COLOSSAL SQUID

There is another species of extremely large squid that's not very closely related to *Architeuthis*, and that's the Colossal Squid (*Mesonychoteuthis hamiltoni*). This species is a type of cranchiid, or glass squid, although at adult sizes it is not transparent but tends to be dark red. This is a popular color partly because it is attenuated quickly and becomes invisible in the mesopelagic, and partly because it is easy for cephalopods to produce with their chromatophores—many shallow nearshore cephalopods, such as the Giant Pacific Octopus (*Enteroctopus dofleini*) and the Pacific Red Octopus (*Octopus rubescens*, pages 104–105), also default to red.

Colossal Squid, while they can swim and have been recorded on video flapping their large round fins, are more likely to be sit-and-wait predators rather than swim-and-catch predators. This is consistent with their general cranchiid morphology and physiology. They get heavier than but not as long as Giant Squid, with the heaviest adult measured at 1,091 pounds (495 kg). Depending on what metric you want to use, length or weight, you can claim either of these two species as the largest squid and the largest cephalopod in the world. Colossal Squid also hold the record for the largest eyes, not only of cephalopods but of all animals, growing up to 16 in (40 cm) in diameter.

→ A Colossal Squid is gaffed alongside a fishing boat in the Southern Ocean. Fishing gaffs are used to bring large fish (or squid) on deck after they've been caught at depth and brought to the surface.

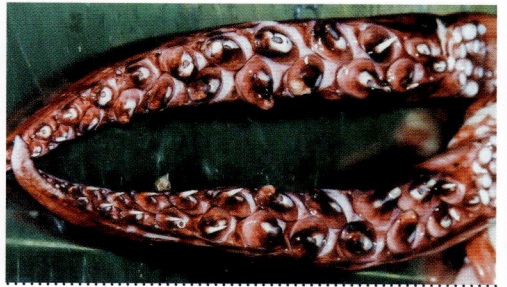

COLOSSAL HOOKS

Not much is known of Colossal Squid biology, but they have additional fascinating morphology: instead of suction cups, both their arms and the clubs of their tentacles bear hooks. These hooks can swivel in place—Colossal Squid are the only cephalopods with such an ability. Possibly, they are so large that they need larger prey, which may be so strong it could break free of mere suction cups or even immobile hooks. The mobility of the hooks could make them more difficult to escape from.

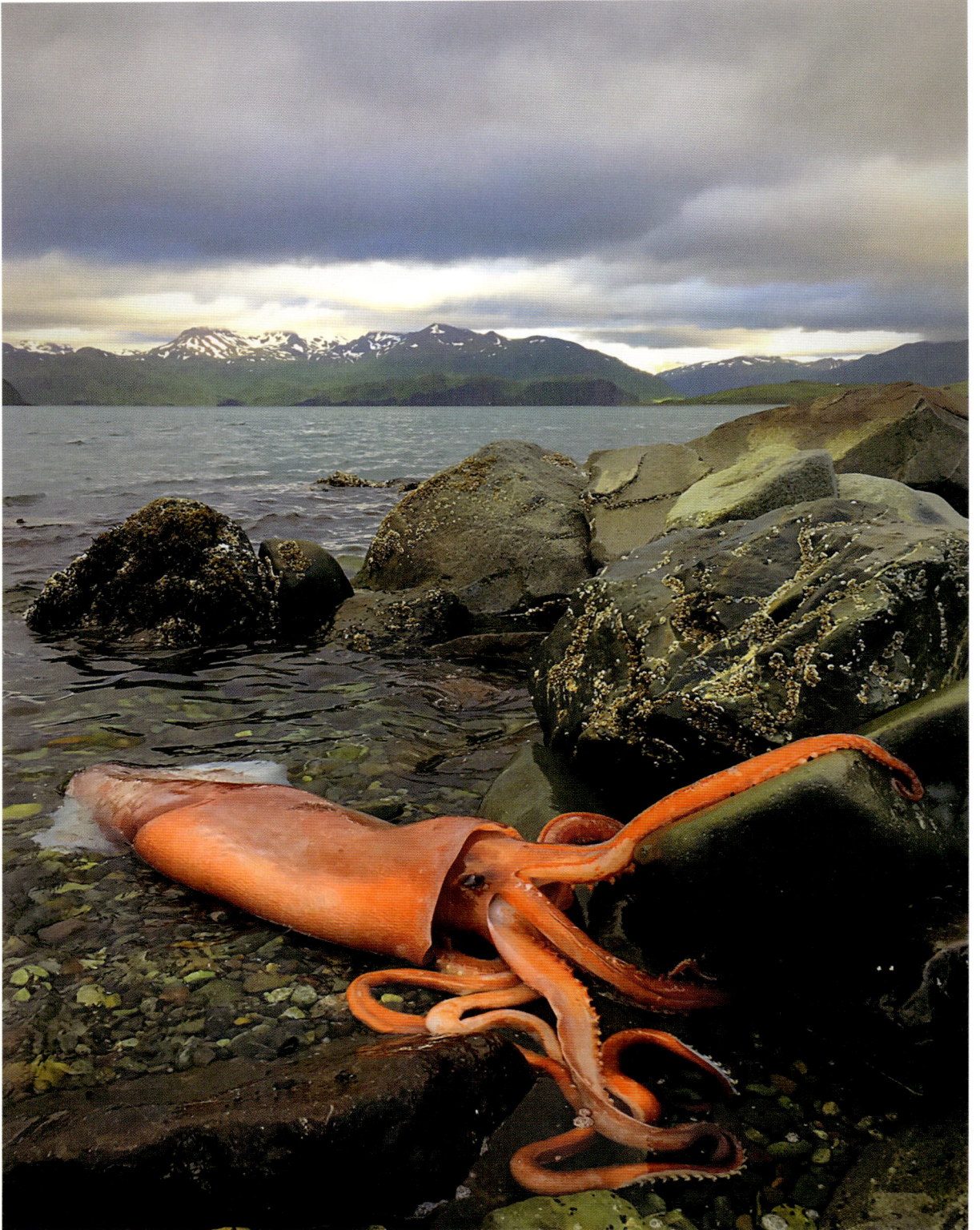

Like other glass squid, Colossal Squid have photophores. They are known to both eat and be eaten by toothfish, species that are also fished by humans. One of the only two videos of live Colossal Squid in the wild was taken when an individual tried to eat a toothfish captured by a fishing vessel. They undergo ontogenetic but not diel migration (see pages 150–153).

ROBUST CLUBHOOK SQUID

Another very substantial species of squid, often included in lists of the biggest cephalopods though it is not known to reach near the lengths of either Giant or Colossal Squid, is the Robust Clubhook Squid (*Onykia robusta*). It has the dubious honor of being the subject of a misleading photograph that was circulated for a while as the first underwater photograph of a Giant Squid. A diver is seen in the background, and if both diver and squid were at the same distance, the squid would be of a truly astonishing size, but it's a perspective trick and the diver is actually quite a bit further away. Furthermore, the species is a Robust Clubhook Squid.

The mantle of Robust Clubhook Squid can be as long as a very tall human, up to 6½ ft (2 m). They belong to an entire family of clubhook squid (Onychoteuthidae) the name of which refers to the fact that the clubs of their tentacles, though not their arms, have hooks rather than suckers. These hooks serve a similar purpose to suckers, catching and holding prey, and it is not known why this particular group evolved hooks. They are believed to have evolved from the chitinous sucker rings possessed by all open ocean squid, which are lined with very small teeth. Over time, one of these teeth may have evolved to be much larger than the others, while the suction cup itself became reduced and then nonexistent.

Robust Clubhook Squid live in the North Pacific from California to Japan. They roam readily into boreal (the subarctic region comprising seas and oceans just south of the Arctic Ocean) but not truly Arctic waters. Paralarvae have been collected near Hawaii, suggesting that they may be tropical breeders. Sperm whales and sleeper sharks eat them. Preliminary genetic research suggests that they are highly migratory, either as adults or as paralarvae or both, without distinct populations. This is similar to our current understanding of the population connectivity of Giant Squid.

← The Robust Clubhook Squid is not an intertidal species; this individual has been carried ashore by waves. It may be senescent, its post-reproductive body breaking down.

Tentacle clubs

Squid have a variety of tentacle clubs. The typical structure is a widened club or manus with several rows of suckers (above), but clubhook squid have several rows of hooks instead (bottom). These tools are not unlike fishing gaffs, hooking prey and bringing it to the mouth.

Tentacle club with suckers

Tentacle club with hooks

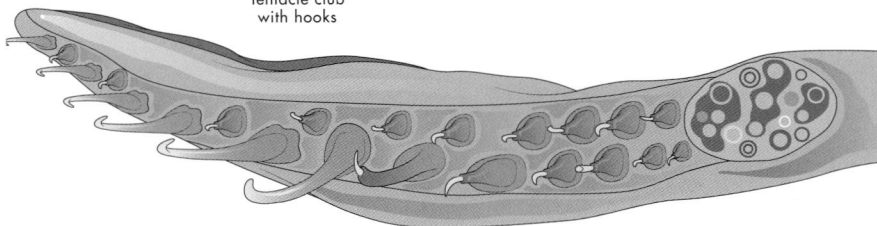

VITRELEDONELLA RICHARDI

Glass Octopus

See-through mystery

SCIENTIFIC NAME	*Vitreledonella richardi*
FAMILY	Amphitretidae
MANTLE LENGTH	4¼ in (11 cm)
TOTAL LENGTH	18 in (45 cm)
NOTABLE ANATOMY	Transparency
MEMORABLE BEHAVIOR	Egg brooding

Unlike squids, octopuses do not have an entire family of glass octopuses, only a single genus, *Vitreledonella*, which is thought to contain a single species. Very little is known about these animals, apart from their transparency. They are believed to brood eggs in their mantle cavity, but the brooding itself has never been observed—simply surmised from the fact that many hatchlings were collected along with a female octopus in 1937.

Like many mesopelagic animals, Glass Octopuses have peculiar eyes to capture the slight bits of available light. Their eyes are compressed into disklike protruding rectangular shapes. The transparency of their bodies is also a clear (ha!) adaptation to the midwater habitat, where it renders them nearly invisible. They are not without chromatophores, however, as was documented in footage taken by the ROV *SuBastian* in 2021. In the video, webs between the Glass Octopus's arms were spotted with vivid yellow dots where chromatophores were expanded. The ROV was exploring the tropical Pacific; Glass Octopuses have been reported from tropical waters all around the world, including the Arabian Sea.

→ The suckers of this species grow in a single row along the entire arm length but they are significantly larger past the arm web. The outer arms are probably used more than the inner arms for grabbing prey, hence the greater size of the suckers.

Both the transparency of the body and the large yellow chromatophores are reminiscent of octopus paralarvae. When scientists looked closely, they found several additional features these animals share with young octopuses of other species: reduced internal organ size, Kölliker bristles (which are thought to aid in hatching), and a toothed beak. They proposed that Glass Octopuses, along with several other closely related mesopelagic octopuses, are neotenous—that they evolved over time from octopuses with more typical adult forms through genetic mutations that caused the paralarval traits to persist into adulthood.

This putative neotenous group includes the Telescope Octopus (*Amphitretus pelagicus*, pages 202–203) and the Diaphanous Pelagic Octopus (*Japetella diaphana*, pages 206–207). When researchers studied their genes, they found that these species are not closely related to other pelagic octopuses, like argonauts and Football Octopuses (*Ocythoe tuberculata*, pages 166–167), but are rather close cousins of benthic octopuses that live on the seafloor. The authors proposed support for a theory dating back to the 1920s, that the ancestors of some pelagic species were benthic adults with planktonic paralarvae. Over time, the theory goes, the paralarvae remained longer and longer in the plankton, eventually developing there all the way to reproductive maturity. The evolutionary history and relationships between groups of octopus species remains an area of active research.

AMPHITRETUS PELAGICUS

Telescope Octopus

Tube-eyed weirdo

SCIENTIFIC NAME	*Amphitretus pelagicus*
FAMILY	Amphitretidae
MANTLE LENGTH	3 ½ in (9 cm)
TOTAL LENGTH	8 in (20 cm)
NOTABLE ANATOMY	Tubular eyes
MEMORABLE BEHAVIOR	Moveable orientation of eyes

This small gelatinous pelagic animal is named and known for its eyes, unique among octopuses. The two eyes are placed very close together, rather than on opposite sides of the head as is typical for cephalopods. They look somewhat like the eyestalks of a crustacean, and the bases touch while the tubes angle away from each other to form a V.

Although the Telescope Octopus hasn't been studied extensively, scientists have been able to collect a living specimen and keep it alive in an aquarium on a research vessel long enough to make some behavioral observations. They saw that it orients the darkest, strongest shadow-casting parts of its body vertically, in order to cast a minimal shadow to any predators or prey below. These are its digestive gland and its two strange eyes. And although it lives an entirely pelagic lifestyle as far as we know, this captive octopus settled and crawled on the bottom and sides of the aquarium.

The breathing, respiration, and metabolic rate of the Telescope Octopus have been studied as part of research into how the metabolism of cephalopods in the midwater changes with depth. The metabolism of the Telescope Octopus fell into the same range as other pelagic octopuses, low enough to be similar to jellyfish, much lower than shallow-water cephalopods and also lower than deep-water squid.

Like many other pelagic octopuses, it has a single row of suckers on its arms, rather than the double row typical of benthic octopuses. Toward the tips of the arms, however, the single row becomes doubled. Intriguingly, it lacks fins as well as any internal shell remnants. It is committed to its gelatinous lifestyle.

Another distinctive feature of this species is that the funnel is completely fused to the mantle, rather than protruding from it as in other cephalopods. It's not always easy to distinguish the mantle from the head in any octopus, but in the Telescope Octopus it is harder than most. Instead of a large gap through which water can be drawn for breathing and jet propulsion, this species has evolved two smaller openings on either side of its head.

The Telescope Octopus lives all around the world in the tropics and subtropics and has been found as deep as 6,500 ft (2,000 m) below the surface.

→   The bluish tint to the gelatinous parts of this octopus is a reminder that blue light penetrates deeper into the ocean than any other colors. The solid shadow-casting nature of the octopus's eyes and digestive gland are also apparent here.

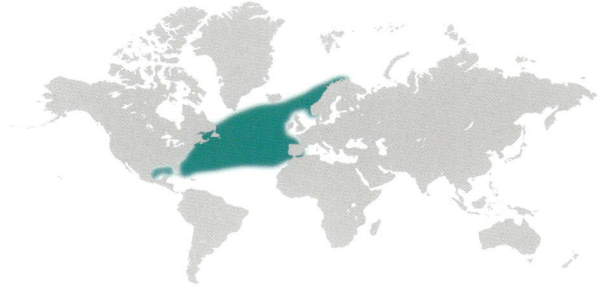

ARCHITEUTHIS DUX

Giant Squid

World's longest cephalopod

SCIENTIFIC NAME	*Architeuthis dux*
FAMILY	Architeuthidae
MANTLE LENGTH	16½ ft (5 m)
TOTAL LENGTH	59 ft (18 m)
NOTABLE ANATOMY	Huge size
MEMORABLE BEHAVIOR	Stranding

Giant Squid have long been known from specimens washed up on beaches or found floating dead or moribund at the surface. It was not until 2006 that video was recorded of Giant Squid in their natural habitat, thanks to the determined efforts of scientist Tsunemi Kubodera.

Sperm whales are the most famous predators of Giant Squid, although any large marine predator will gladly eat them if it can get a bite, including seabirds such as albatrosses. Sperm whales, however, are the only predator that can regularly dive deep enough to hunt for adults in their natural habitat, rather than waiting for dying animals to rise to the surface.

Although an actual encounter between a sperm whale and a Giant Squid has never been recorded, we can surmise a great deal about it from circumstantial evidence. For starters, sperm whale stomachs are full of beaks, and rough calculations based on the number of beaks indicate that Giant Squid are an extraordinarily abundant resource, with approximately 2 million individuals being eaten by sperm whales every day—meaning the entire population must be much larger.

Next, no piece of a sperm whale has ever been found inside a Giant Squid stomach. So although the interaction might be framed as a "battle," in reality, it is a case of a predator (sperm whale) attacking its prey (Giant Squid). The prey can sometimes damage its predator (just like an

antelope can kick a lion), which is why we do see the marks of suction cups on the skin of sperm whales and sharks.

It's been thought at times that there were many species of Giant Squid, but current genetic studies indicate they all belong to the same global species, *Architeuthis dux*. Because they are so large already and humans are prone to hyperbole, lots of "fish stories" have made their size even greater. The maximum known size, however, is 43 ft (13 m) in total length and 7 ft 3 in (2.2 m) in mantle length. Their arms and tentacles are far longer than their mantles, which greatly inflates their "total length" over their "mantle length."

More Giant Squid have been viewed in recent years, thanks to advances in underwater exploration, as well as increases in deep-sea trawling. It may also be due to climate change shifting water circulation and currents so that more individuals are brought into warm shallow water.

→ This famous photograph of two scientists, Erling Sivertsen and Svein Haftorn, measuring a Giant Squid was taken in Norway in 1954. Ten years earlier, Sivertsen had been arrested during the German occupation of Norway, and might not have survived to see this squid if the Allied victory had come later or not at all.

Diaphanous Pelagic Octopus

On-again-off-again octopus

SCIENTIFIC NAME	*Japetella diaphana*
FAMILY	Amphitretidae
MANTLE LENGTH	2½ in (6 cm)
TOTAL LENGTH	4½ in (12 cm)
NOTABLE ANATOMY	Oral photophore ring in females
MEMORABLE BEHAVIOR	Switching between color and transparency

Like other members of its family, Amphitretidae, the Diaphanous Pelagic Octopus broods eggs in its arms. There seems to be a distinct difference between pelagic brooding octopuses, which hold all their eggs fully contained in their arms (or shell, in the case of argonauts), and pelagic brooding squid, which trail an egg mass or egg sheet longer than the length of their own body.

Females are thought to brood just below the reach of sunlight, at around 2,600 ft (800 m), for safety in the darkness. Hatchlings are found nearer to the surface, so it is concluded that they either swim up after hatching or are brought to that environment by the mother. They undergo ontogenetic migration as they grow, moving gradually deeper until they arrive at mating and brooding depth. However, they do not seem to engage in diel migration, remaining at the depth typical of their age throughout the day–night cycle.

Sexually mature females have a striking circular yellow photophore around the mouth at the base of the arms, which could be used as a signal to males of mating readiness.

→ Males of this species do not possess a hectocotylus, a modified arm for sperm delivery. They may deliver sperm with a non-hectocotylized arm, or with a terminal organ similar to a penis.

Photophores are generally much rarer in octopuses than in squid, which makes this feature especially noteworthy. Diaphanous Pelagic Octopuses can be either mostly transparent or completely yellow–orange. Both of these can be protective, since some predators have difficulty seeing a transparent organism, while other predators use their own photophores as searchlights to bounce off any reflective surfaces inside a transparent organism. The octopus's expanded chromatophores are opaque to such searching techniques.

The Diaphanous Pelagic Octopus's global distribution is restricted to tropics and subtropics except in the North Pacific, where it goes into chillier northern waters. This may be due to northern expansion of the oxygen minimum zone, where it dwells. These animals don't require OMZs, though, which means that scientists can compare those individuals that do live in OMZs with those that don't to see the differences between them. It turns out that both have low metabolic rates, yet individuals that live in OMZs are better able to handle low oxygen, signifying that they have additional adaptations beyond their metabolic rate. Unlike many other marine species, they're better able to handle low oxygen the higher the temperature gets.

They can't live in the very least oxygenated part of the OMZs, like many other OMZ-associated species that use the region but remain on the edges, where it's a little easier to breath. This creates a fairly compressed habitat, where everyone's crowded into a narrow space—probably a great place for a predator like an octopus to find food.

Dana's Chiroteuthid Squid

Fancy-tailed imitator

SCIENTIFIC NAME	*Planctoteuthis danae*
FAMILY	Chiroteuthidae
MANTLE LENGTH	2¼ in (5.5 cm)
TOTAL LENGTH	3½ in (9 cm)
NOTABLE ANATOMY	Long elaborate tail
MEMORABLE BEHAVIOR	Creating a tentacle sheath with the arms

As a member of the Chiroteuthidae family, this species has the distinctive doratopsis paralarva. The stiff gladius, or internal shell, which runs the length of the mantle in other squid, extends far beyond the mantle tip and bears complicated decorations that are thought to mimic the appearance of a stinging siphonophore. Doratopsis paralarvae also have an elongated "neck" separating the head and the mantle.

The genus *Planctoteuthis* is unique within the chiroteuthids in that the long tail of the doratopsis is retained into adulthood. It's thought to have evolved by the process of neoteny, in which juvenile traits previously lost are kept for longer and longer because they prove advantageous.

Although the tail is this species' most obvious and striking feature, its arms are also unusual. Two of the arms, which have very few suckers, can be "zipped up" along their length to form a sheath for the two tentacles (the appendages that can be extended and retracted elastically in most squid). In *Planctoteuthis*, the tentacles appear more inclined to coil than to retract. Since arms and tentacles are used for catching prey, it seems likely that this has to do with the diet of these squid, but it hasn't been studied yet.

The females of this species have greatly enlarged suckers on four of their arms, looking rather similar to the enlarged suckers on the arms of male Deep-sea Bobtail Squid (*Heteroteuthis hawaiiensis*). Since in both cases the suckers are sex-specific, it seems likely that they are related to mating or courtship. Is it possible that in one species males must attract females, while in the other, it's the other way around? This remains pure speculation, as recordings of these animals in the deep sea have not yet provided data on their reproductive habits.

Until ROVs and underwater cameras made it possible to document these squid *in situ*, the true extent of their remarkable tails was unknown, since they are too fragile to withstand capture by net. They are never found in surface waters, but as paralarvae inhabit the midwater, where visual predators would notice and be dissuaded by their camouflage. As they age, they move into deeper waters, well below the limits of where light reaches, but continue to rise to shallower waters at night. This diel migration may be why it remains beneficial for them to retain their siphonophore mimicry into adulthood.

→ The adornments on the long "tail" of this squid are thought to imitate siphonophores, a colonial relative of jellyfish that can have extremely elongate "stems" carrying various floating, reproductive, and feeding polyps.

Piglet Squid
Blue-ribbon funnel

SCIENTIFIC NAME	*Helicocranchia pfefferi*
FAMILY	Cranchiidae
MANTLE LENGTH	4 in (10 cm)
TOTAL LENGTH	4½ in (12 cm)
NOTABLE ANATOMY	Very large funnel
MEMORABLE BEHAVIOR	Losing tentacles with maturity

The Piglet Squid is so named because of its appearance when viewed head-on, since its large funnel can resemble a pig's snout. This species is a type of glass squid, and like the others of this group, it undergoes both ontogenetic migration, moving into deeper water as it grows older, and diel migration, traveling up at night and down during the day.

Found throughout the global ocean in tropical and subtropical waters, the Piglet Squid was first described in 1907 from the coast of Ireland. At this time its large funnel came to the attention of science, but the "Piglet Squid" moniker was not given until a century later. The species name honors Georg Johann Pfeffer (1854–1931), a German naturalist specializing in cephalopods. The large funnel is present from hatching, making this species a relief to oceanographers who often have great difficulty identifying paralarval squid species.

Piglet Squid fins are almost comically small, resembling the tiny toupee-style fins of other species' paralarvae. This reminds one of the theory of a possible neotenous origin for certain pelagic octopuses. The authors of that theory hypothesized that glass squid might have a similar neotenous origin. Certainly, part of the charm of the Piglet Squid is its resemblance to a baby. Curiously, its tiny fins do not attach to the mantle proper, but to an extension of the gladius (the stiff internal rod that evolved from the shell of ancestral cephalopods to be an internal support structure for squids) that extends slightly beyond the tip of the mantle. A similar, but much more dramatic extension of the gladius is found in the paralarvae of other oceanic squid and in both paralarvae and adults of Dana's Chiroteuthid Squid (*Planctoteuthis danae*, pages 208–209).

Although, like other glass squid, Piglet Squid are translucent to transparent, they are covered with a light dusting of chromatophores and become redder as they age. Red, as you may recall, is invisible in the deep sea, so this contributes to their camouflage, and is probably connected with the increased depth to which they descend. Another curious feature of aging in this species is the loss of the two long tentacles, so that in mature adults often only eight arms are present. This loss appears to be common in other glass squid as well, though the reason for it is not known.

→ The bright line inside this Piglet Squid comprises the animal's digestive gland and ink sac, the only internal organs that cannot be made transparent. The squid orients them vertically to cast the least shadow.

DEEP SEA

The midnight zone

The deep sea isn't quite a stranger at this point, because many of the species we've encountered in previous chapters spend part of their lives in the bathypelagic region. In this chapter, we'll touch on the bathypelagic and also venture beyond into abyssal and hadal regions, as well as the benthic deep sea—the habitat on the deep seafloor. In some places the "deep" seafloor is only a few thousand feet deep; in other places it is many thousands.

THE SHAPE OF THE DEEP

The deep seafloor is not a uniform sandy or muddy habitat, any more than the seafloor in nearshore environments. As mountains rise from dry land, so mountains rise underwater. When they break the surface, we call them islands; when their peaks remain submerged, they are seamounts. Many undersea geologic monuments are not single mountains but long chains like mountain ranges on land, and these are known as submarine ridges. There are also volcanoes, rifts, valleys, and trenches to explore.

The bathypelagic stretches from 3,300 to 13,100 ft (1,000–4,000 m), and in some places, around continents and seamounts and ridges, the bathypelagic reaches the seafloor and there are no more layers. But where the seafloor is deeper, you find the abyssal region, from 13,100 to 19,700 ft (4,000–6,000 m), and the hadal region, which is everything below that.

The deepest places on Earth are ocean trenches. The Mariana Trench is the deepest of these, containing the planet's deepest known location: the Challenger Deep, so named because it was first plumbed by the

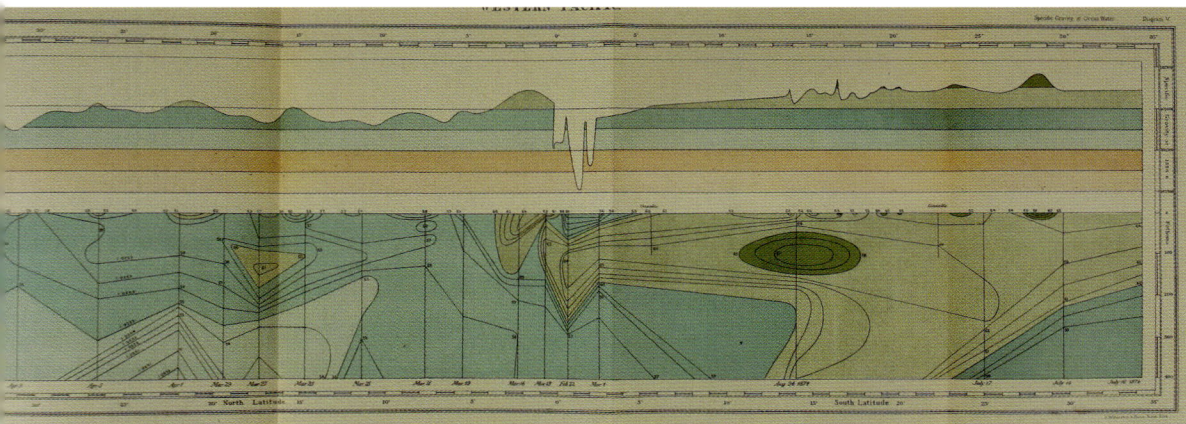

ship HMS *Challenger* in 1875. It's more than 32,800 ft (10,000 m) deep! We do not know of any cephalopods living at this depth, but that doesn't mean there couldn't be any. We simply don't spend a lot of time exploring trenches, so we don't have a thorough sense of what can live there.

DEEP-SEA RECORDS

Cephalopods are especially tricky to collect at extreme depth because of their mobility. Rarely, octopuses have been caught in deep-sea trawl nets, but such nets sample a range of depths, so they could scoop an octopus from shallower water on the way up or down. Before the 2020s, although abyssal cephalopods were known, there were no definitive records of cephalopods from the hadal zone. However, in 2020 and 2022 respectively, further exploration pushed the deepest record of an octopus to 4.3 miles (7 km) and the deepest record of a squid to 3.9 miles (6.2 km)—both measurements unequivocally in the hadal region.

The octopus was a species of *Grimpoteuthis*, or dumbo octopus, recorded in the Java Trench by a "lander," a device that scientists drop to one position on the seafloor to record video for a while and then recover. (Quite similar to the landers we send to other planets, in fact—one of many similarities between deep-sea and space exploration.) The squid was a *Magnapinna*, or bigfin squid, recorded in the Philippine Trench by an HOV.

↗   These may appear to be "deep-sea divers," but the sunlight filtering from above shows they are not in the bathyal zone. The deepest recorded scuba dive reached just over 1000 feet (300 meters).

→   The Dumbo Octopus is an ambassador for deep-sea cephalopods, with its large ear-like fins and soft cuddly-looking body.

←   Scientists on the *Challenger* expedition measured many physical, chemical, and biological features of the ocean on a large scale for the first time. Specific gravity, a proxy for salinity, is graphed here against latitude.

HISTORY OF DEEP-SEA EXPLORATION

The *Challenger* expedition marked the first time humans set out with the express purpose of deep-sea exploration, and is famous for discovering thousands of new species as well as disproving the hypothesis that the deep sea was full of living fossils and protoplasmic goo. Even though there were no submersibles at the time, the *Challenger*'s nets brought up many fascinating life-forms that were not "primitive" or "missing links."

Leaping forward a century and a half, the new deepest records for octopus and squid both came from a global deep-sea exploration project mounted by independently wealthy explorer Victor Vescovo, with chief scientist Alan Jamieson. This Five Deeps expedition, carried out from 2018 to 2019, brought humans to the deepest points of all five oceans (Pacific, Atlantic, Indian, Southern, Arctic) and conducted 100 lander drops. They collected an enormous amount of video and other scientific data, much of which is still being analyzed—which is why the "discoveries" of the cephalopod depth records lagged behind the expedition itself.

Humans have been fascinated with the depths of the sea for longer than recorded history, no doubt—just like we're fascinated with outer space. But the deeps are in some ways better, because we *know* there are aliens down there. We just have to keep looking, and we're bound to see new and fascinating animals. For example, a new species of octopus, the Emperor Dumbo (*Grimpoteuthis imperator*), was just described in 2021 from near the Aleutian Islands.

↖ *Limiting Factor*, the HOV of the Five Deeps Expedition, followed and built on the legacy of Project Nekton's *Trieste*, which took two humans to the Challenger Deep in 1960.

→ This dead whale must have fallen to the seafloor recently, since large scavengers have not yet converged to consume its flesh. Microscopic scavengers are no doubt already at work.

SINKING FOOD

Where sun-loving algae cannot survive, either independently or within coral hosts, what sustains food webs? For most parts of the deep sea, the base of the ecosystem is still photosynthesis in surface waters. We know that much of what sinks is consumed in the OMZ, leaving only a few scraps to fall beyond those waters, which is one reason that many animals inhabit both midwater and bathypelagic depths. However, the animals that live in the OMZ also molt and defecate and die, making their remains available to ever deeper scavengers.

From time to time, much bigger sinkers, such as dead whales, arrive on the seafloor. A whale fall constitutes an entire ecosystem, a buffet blessing gifted from above. A single dead whale can support a huge variety of life for decades. First, large scavengers such as sharks and hagfish arrive, devouring the flesh while it's still present. Smaller scavengers such as crabs and octopuses also join the feast. Eventually, the community shifts to specialists that target the bones, with an entire group of deep-sea worms known as bone worms for their ability to colonize a skeleton. Eventually, bacteria take over the decomposition process, producing bacterial mats that themselves support a new layer of grazing organisms including mussels and snails.

Bone worms, and other whale fall specialists, may have evolved long before modern whales. Evidence suggests that large marine animals have been sinking to the deep seafloor since dinosaur times. Marine reptiles such as plesiosaurs and pliosaurs and mosasaurs would have provided extensive sustenance. Scientists have even experimented with dropping dead alligators to the seafloor to study the successional communities that arrive to take advantage of the nutritional bonanza. However, there are also ecosystems in the deep sea that do not depend on food falling from above.

Hot vents and cold seeps

The deep seafloor was once thought to be a dead place, too dark and cold to support life. Ever since early expeditions in the mid 1800s dredged life from the depths and disproved this theory, every new deep-sea exploration has revealed more and more surprises.

→ This geyser erupting in Kamchatka, Russia, represents the type of hydrothermal activity humans are familiar with on land.

↓ Named "Champagne Vent" for its bubbly nature, this white smoker was found near the Mariana Trench.

HYDROTHERMAL ECOSYSTEMS

One of the greatest surprises came in 1977 with the discovery of hydrothermal vents. These are places of volcanic activity, but instead of the classic explosive volcanoes we're used to, vents are continuous leaks of heat and chemicals from inside Earth. The existence of these vents significantly changes two of the things about the deep sea that made people think of it as antithetical to life. They are a kind of "deep-sea sun." Near the surface, the sun provides both requisite warmth and the necessary energy to make food. In the depths, hydrothermal vents provide less welcome warmth, but also a source of energy: chemicals that can be used to produce food just like sunlight. Entire ecosystems can be built not on photosynthesis, but on chemosynthesis.

GEOTHERMAL HEAT

The solid crust of Earth is a (relatively) thin layer of plates sliding on molten rock beneath. The heat inside Earth that keeps this rock melted does not come from the sun, but from the planet itself. Much of that heat is left over from when Earth and the rest of the solar system were formed. As gravity gathered space dust and rocks into clusters, and larger and larger chunks began to slam together, these collisions produced a great deal of heat. Most of it was kept inside the growing Earth. More heat is still being produced inside the planet by the friction of moving materials rubbing together, as well as by the decay of radioactive elements.

Of course, heat is also being lost, and eventually Earth will bleed out all of its heat and become a cold, dead rock like many other planets. But that's not going to happen for an extremely long time, because the crust is an effective blanket. So the inside of Earth holds onto its heat, and places where the crust splits or breaks reveal just how smoking hot it is.

Plate tectonics refers to the constant, if very slow, movement of the plates of Earth's crust. In some places, plate edges push together. Where two continental land masses push together, they uplift into mountains. Where an ocean basin pushes into a continent, it moves beneath, often also raising mountains as it goes, in addition to producing trenches, volcanoes, and earthquakes. The same can happen where two ocean basins push together:

trenches and volcanoes. In other places, plate edges are moving apart. This is happening near the center of the Atlantic and Pacific Oceans, expanding both ocean basins as new material rises into the open space to create an aquatic mountain habitat as well as more volcanoes and earthquakes.

On land, we're familiar with volcanoes as locations where the melted rock, or magma, inside Earth bursts free as lava. It can happen in a slow trickle or an explosive flood, and the latter can throw enough particulate matter into the air to alter the climate. This phenomenon is thought to have contributed to most of Earth's major mass extinctions in the past. But what about underwater volcanoes?

Black smokers vs cold seeps

Cold seeps and hot vents release the same chemicals into the water and support many of the same animals. However, cold seeps last longer than hot vents and can occur separately from volcanic activity.

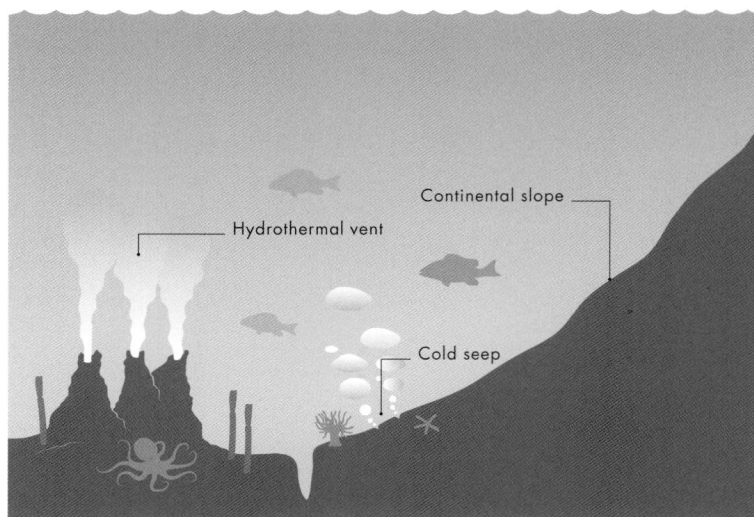

TAXONOMY OF UNDERWATER VOLCANOES

What happens when a crack opens in the seafloor? Water immediately flows down into it. As it flows along, the seawater's chemistry is changed. It also picks up Earth's body heat along with a variety of dissolved chemicals, and the hotter it gets, the more it expands and seeks an escape. Hydrothermal vents are a bit like underwater teakettles going off.

However, they don't literally produce steam—the water remains liquid, rather than turning into gas. That's because of the great pressure at depth, which can prevent water as hot as 700 °F (370 °C) from undergoing a phase transition. Videos of the vents can be puzzling, because the locations *look* like they're steaming. They've even been nicknamed black smokers and white smokers. What we're actually seeing, however, is not gas but billows of superheated water, colored by their chemical composition, emerging into colder water.

Often forming along the ridges where seafloors are spreading, black smokers are the youngest form of hydrothermal vent. The color of the billowing water is produced by sulfur-based chemicals, many of which precipitate out on contact with regular cold deep-sea

water, producing chimneys. Sulfur bound with metals such as iron, copper, and zinc contributes these metals to the nutrients in the deep sea that will be brought to the surface by upwelling, and also creates ores that humans can mine (see pages 224–227).

White smokers are usually older and cooler vents, as the magma beneath the surface hardens and the local heat source fades. Hydrothermal vents are inherently transient, each vent existing for only a few years, but they are also inevitable, with new ones popping up as old ones fade away. The animals that live on these vents must have ways of dispersing from one to another, so their children can colonize a new habitat when the old one dies out.

LIFE FINDS A WAY

Instead of photosynthesis, making energy from light, life down here revolves around chemosynthesis, making energy from chemicals. Bacteria are chemosynthetic experts, and they don't limit such activities to the deep sea—there are chemosynthetic bacteria all over the planet, including at hot springs and in brine pools on land. But it's at the hydrothermal vents where they really shine, supporting an entire ecosystem

COLD SEEPS

"Cold seeps" are locations where chemical-rich water escapes from the seafloor without the heat of volcanic activity. They tend to last longer than hot vents while providing similar ecosystem support materials, including sulfur and methane. Many of the same bacteria-dependent animals thrive at cold seeps as at hydrothermal vents.

as its sole primary producers. The two chemicals most used to produce energy are sulfides and methane. The classic chemosynthetic reaction that hydrothermal vent bacteria employ is combining hydrogen sulfide with carbon dioxide and oxygen to produce sugars and water (just like photosynthesis) with solid sulfur as a by-product.

Chemosynthesis

Like photosynthesizing plants, chemosynthesizing algae "fix" the carbon from carbon dioxide into carbohydrates. But instead of sunlight, they use the energy from breaking down hydrogen sulfide.

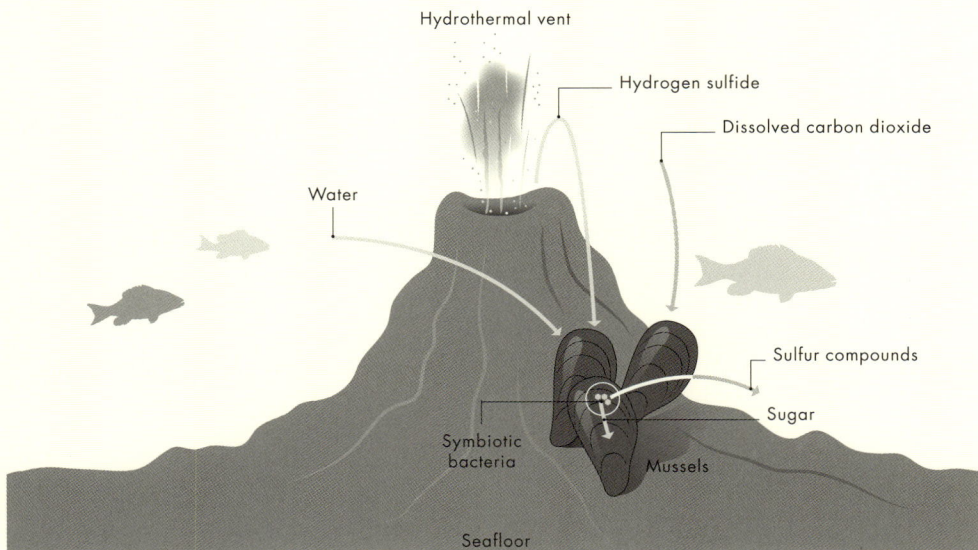

Hydrothermal vent

Hydrogen sulfide

Dissolved carbon dioxide

Water

Sulfur compounds

Sugar

Symbiotic bacteria

Mussels

Seafloor

These vent tube worms grow on a black smoker, their plumes packed with bacteria that can make food from chemicals. At their largest, they can be longer than a person is tall.

Although some snailfish live in shallow water, many are specially adapted to the deep sea, their bodies built to withstand pressure a thousand times greater than at the surface.

Bacteria thrive in the sulfide-rich waters around vents, forming substantial mats. Many small animals such as shrimp graze on the mats, and many larger animals such as octopuses prey on these grazers. There are also animals that have entered into a more intimate association with chemosynthetic bacteria, such as the iconic tube worms, *Riftia pachyptila*. Although their distribution is restricted to the Pacific Ocean, their deep red plumes emerging from tall white tubes are strongly associated with vent life. Rather like the relationship between corals and zooxanthellae, vent tube worms house bacteria inside their bodies, bringing in the raw materials the bacteria need and offering them shelter in exchange for the products of chemosynthesis. Unlike corals, however, tube worms cannot live without their symbionts, since they have evolved away their ability to eat and don't even have a gut. Several species of deep-sea clams and mussels have been discovered engaging in similar symbioses.

What about the scorching temperatures? Well, the incredible heat of the smoker itself drops off precipitously as it mixes with the vast amount of very cold water at the bottom of the sea. Vent animals do not live in and cannot survive the absolute hottest water. However, they do have many astonishing adaptations to survive in much warmer water than their relatives. Pompeii worms have been discovered to tolerate 176 °F (80 °C).

Many scientists think that hydrothermal vents may have been where life first originated on Earth.

We know that some kind of chemosynthesis predated photosynthesis, and the deep sea would have been a more sheltered environment than the surface of Earth back when the planet was being bombarded with meteors in the early days of the solar system.

TRENCHES

Hydrothermal vents, while definitely deep–sea habitats, are as far as we know limited to depths less than 16,400 ft (5,000 m), and are most often found from 6,500 to 9,800 ft (2,000–3,000 m). Incredibly deep, yes, but not abyssal or hadal. The really extreme ocean depths occur in trenches, which are formed where one plate slides beneath another.

The entire hadal zone is trench habitat, which makes it a very small fraction of total ocean habitat. Little food is produced in trenches, but because they are wider at the top and become narrower as they descend, they serve as funnels for food sinking from above.

There is not a huge diversity of animals living in trenches, and those that do tend to have unique adaptations to high pressure. Shrimp have been found, and a particular group of fish known as snailfish that are specially adapted to extreme habitats. Until a few years ago, no cephalopods were known from the hadal zone in trenches, but then exploration revealed new records for both octopuses and squid.

CHEMOSYNTHESIS FOOD WEB

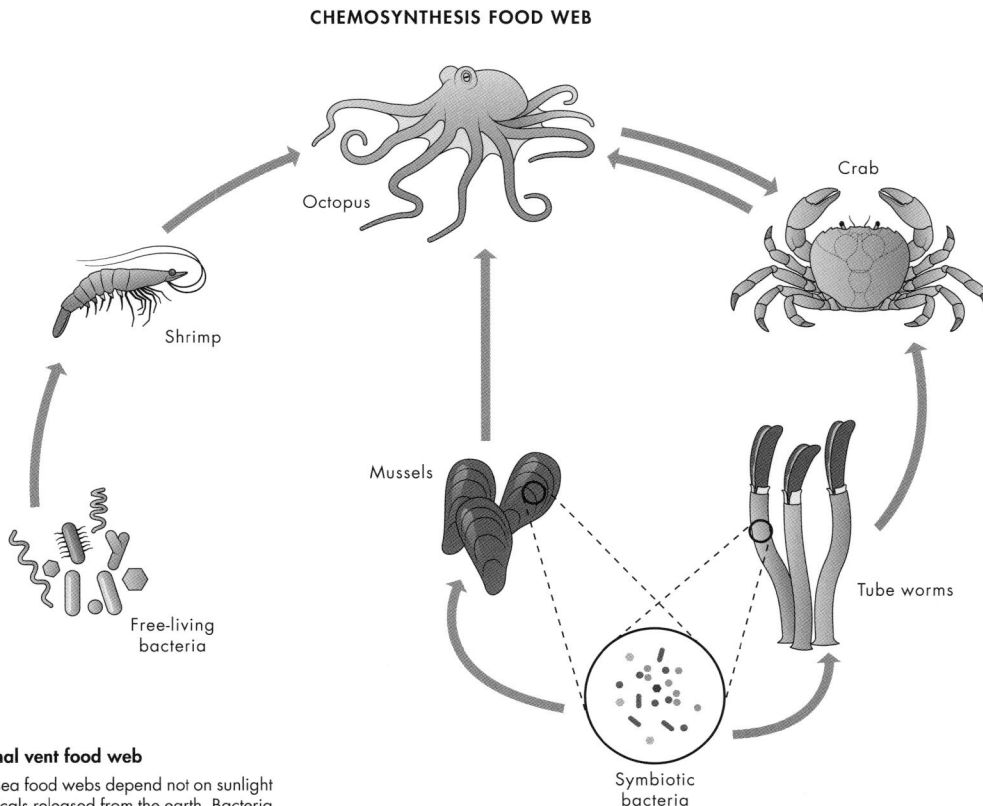

Octopus

Crab

Shrimp

Mussels

Tube worms

Free-living bacteria

Symbiotic bacteria

Hydrothermal vent food web

Many deep-sea food webs depend not on sunlight but on chemicals released from the earth. Bacteria use these to produce sugars, grow, and multiply, and primary consumers either farm or prey on the bacteria. Secondary consumers like crabs eat the farmers, and tertiary consumers like octopuses eat the crabs.

Resource extraction

The minerals brought up by hydrothermal activity can precipitate out of seawater in large quantities, creating mineral ore deposits that are highly valued and desired by humans. These include familiar metals such as copper and gold, as well as rare earth metals with a wide range of uses from electronic components to agriculture. Rare earth metals are especially necessary for the batteries and motors involved in "green" or renewable energy technology, such as solar panels and wind farms. This creates a catch-22, in which moving away from the environmental devastation of burning fossil fuels seems to require moving toward a different kind of environmental destruction.

WEALTH IN THE DEEP

Rare earth metals are not rare in the typical sense—they're found in many different places—but they're generally mixed with other minerals rather than aggregating into pure veins and nuggets like, for example, gold. Rare earth mining operations therefore have to extract metals that make up less than 1 percent of the mineral composition of the deposit. Thus, mining these metals from the deep sea is complicated and destructive, since a large quantity of material must be processed to obtain a relatively small quantity of the desired minerals.

In addition to the metal-containing sulfide deposits of hydrothermal vents, another mining temptation in the deep sea is manganese nodules. These occur in deeper water than most hydrothermal vents, abyssal rather than bathyal. Their size and shape range across what domestic dog owners are used to scooping up on their walks, but manganese nodules are produced

on a vastly slower time frame than dog poop, taking millions of years to grow half an inch. Because of this incredibly slow growth, we're not sure what causes their formation, but it's probably at least partly precipitation from supersaturated seawater. Manganese, their primary component, is used in alloys such as stainless steel and other industrial applications, but nickel is the metal most sought after from these nodules.

PLANS FOR EXTRACTION

Industrial mining of the deep sea is still in the planning and development phase, with no large-scale operations currently underway (as of 2022) and lots of argument about both feasibility and ethics. Numerous companies have acquired permits to conduct mining tests, however, with a pilot test collecting sulfide deposits in 2017 in the Okinawa Trough, and robotic nodule collectors scheduled for deployment in 2022.

Processing is necessary to extract the desired minerals, and it's not in any way feasible to do that *in situ* on the deep seafloor. Nor is it very practical to bring a large quantity of material all the way back to land for processing, so deep-sea mining would involve collecting material from the depths with robots or suction, bringing it up to a processing ship where the valuable bits are taken out, then dumping everything left over (the tailings) back into the water. Both the physical removal of material from the deep seafloor and the return of tailings have the potential to cause significant environmental damage.

225

IMPACTS ON LIFE

Like the growth of manganese nodules, the growth of life in the deep sea tends to proceed slowly. And unfortunately, any region of the deep that's interesting to humans for its resources is also interesting to many other animals. Although deep-sea octopuses have no use for rare earth metals, the same locations that concentrate these metals also concentrate their options for food and shelter. Even manganese nodules, which do not support such an array of symbiotic life as hydrothermal vents, still provide structure in an otherwise featureless environment. Structures offer hiding places (although there is no ambient light to block, prey can still hide from predators' sense of touch and sound), mate-encountering opportunities, and egg-laying locations. Removing these structures will both directly impact life (since mining is bound to kill a number of organisms) and indirectly remove habitat.

The tailings, however, cause some scientists even greater concern. Metals and other minerals that were locked up in the deposits are returned to the sea in diffuse form, and many of these can be toxic. Not to mention the sheer volume of dispersed loose material, which will form a cloud of sediment that could clog the gills and filter-feeding apparatus of animals in the area. If tailings are discarded at the surface rather than pumped all the way back down to the deep, which is a possibility, then they could have even greater impact by blocking the access of photosynthetic organisms to sufficient sunlight.

↖ This deepsea skate lays its eggs at hydrothermal vents. The warmer water causes the embryos to develop faster.

← An octopus explores Davidson Seamount, a habitat off the California coast that in 2008 was added to the National Marine Sanctuary to protect its profusion of life.

↗ The Ghost Octopus lays eggs on sponges that live on manganese nodules, a cascading dependency on structures that would be affected by deep-sea mining.

→ This octopus nestles in a crevice on a canyon wall. Even the simplest physical structure can be important for survival.

CEPHALOPODS, SPECIFICALLY

A new species of octopus colloquially called the Ghost Octopus may be at risk from human ventures even before getting a proper scientific name. Ghost Octopuses received their common name due to their apparent lack of chromatophores, a feature they share with the Deep-sea Vent Octopus (*Vulcanoctopus hydrothermalis*, pages 236–237). The Deep-sea Vent Octopus also lacks an ink sac, which may or may not be true of the Ghost Octopus. Some midwater octopuses also have lost or reduced chromatophores and ink sacs. However, midwater octopuses have switched out their abundant chromatophores for transparency, as in the Glass Octopus (*Vitreledonella richardi*, pages 200–201), because there's still enough light in their environment, filtering faintly from above or directed at them by the luminous searchlights of their predators, that they benefit from reflecting no light at all. Meanwhile, both Ghost and Deep-sea Vent Octopuses are simply white. There is so very little light in their environment that it doesn't matter what color they are. Both have eyes; however, ROV lights, which would be blindingly bright at these depths, don't seem to trigger a reaction in Deep-Sea Vent Octopuses, so the extent to which they can actually see is unknown.

Octopuses are more mobile than many of the species that live at hydrothermal vents and other potential mining sites, so in some ways they may be more able to escape the impacts. However, their reproduction and development are weak points. Deep-sea Vent Octopuses seem to live sexually segregated, with scientists discovering only males for years before the first female was found. That suggests they may have timed behaviors to meet for mating, which could be easily disrupted by mining. Meanwhile, Ghost Octopuses lay their eggs on sponges on the seafloor, a vulnerable location that they can't pick up and move if a mining robot comes along.

Visual communication in the dark

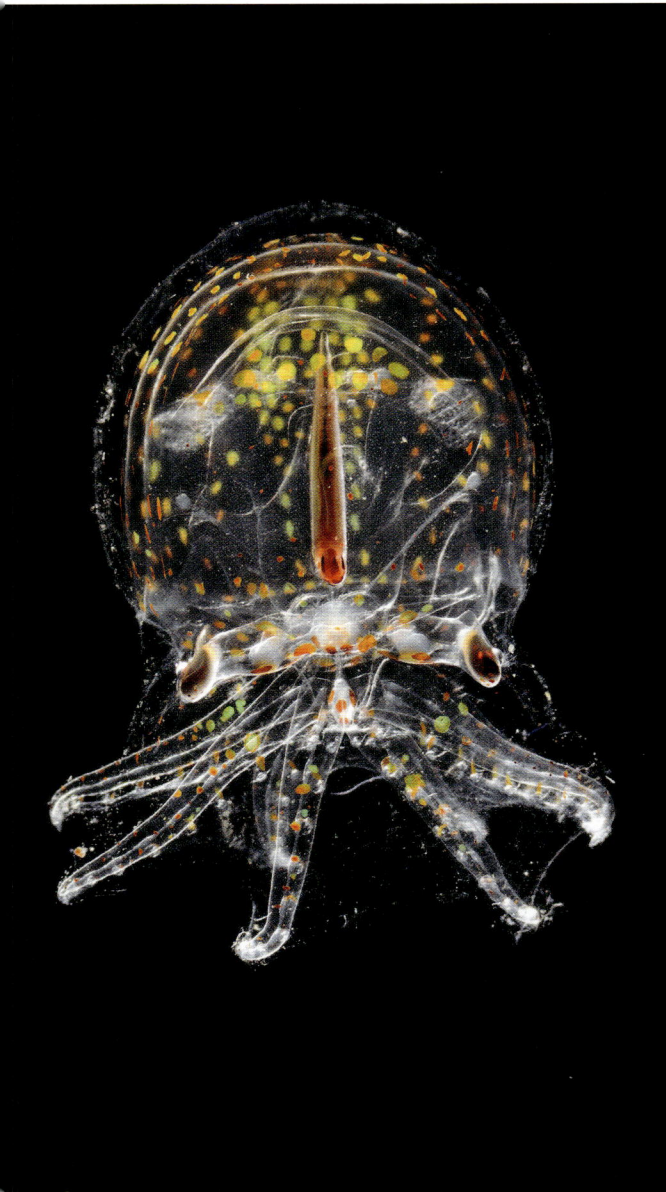

One reason that cephalopods captivate us humans is that they are such visual and colorful animals. Most humans, too, depend on sight more than any other sense, and are captivated by visually arresting organisms in nature. But visual possibilities decline with increasing depth in the sea; how do cephalopods adapt?

CHROMATOPHORES AND PHOTOPHORES IN THE DEEP

In nearshore or epipelagic waters, abundant displays of chromatophores, iridophores, and leucophores are used for both camouflage and communication. As the water becomes deeper, and red is completely eliminated, camouflage takes on new possibilities: cephalopods in these waters are often close to transparent, or colored red or black. Deeper still, as we saw in the previous chapter, transparency and red coloration become nearly ubiquitous (think of the Glass Octopus, *Vitreledonella richardi* (pages 200–201) and the Vampire Squid, *Vampyroteuthis infernalis*) and counter-illumination is crucial—here photophores are often used to erase any shadow that might otherwise be cast.

← Unlike in the skin of shallow-water octopuses, the sparse chromatophores of this Glass Octopus show up as individual spots, along with the animal's red digestive organ, curved internal shell, and cloudy central nervous system.

INK TECHNIQUES

Many deep-sea squid have and use ink, although at first we might expect a smokescreen to be useless in the deep dark. Deep-sea octopuses have mostly lost their ink, but at least 17 species of squid have been observed releasing ink in the deep. Even though it's dark, ink may still serve to block what little light there is, and it also has a chemical component, capable of confusing predators' noses or equivalent chemosensory anatomy. Small puffs of ink could even act as chemical communication between squid of the same species, when it is too dark to communicate by skin patterns. Furthermore, some species of glass squid were observed releasing ink inside their own transparent body, turning it opaque. This may be a way for them to switch between transparency and black coloration when the situation calls for it, as they move up and down in the water column or experience predators with different hunting strategies.

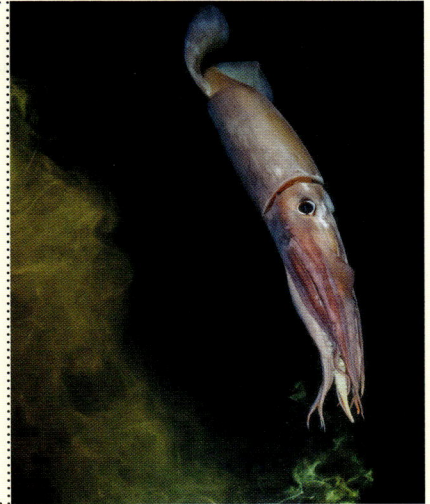

Photophores, or light organs, are the rule rather than the exception in deep-sea squid. Most species have one or more photophores to counteract the darkness of their eyes and digestive organs. Photophores can also be found on the tips of the arms, or in lines along the arms or body. These could be used for communication with other members of their species, or perhaps to lure or confuse prey.

However, characteristic chromatophore patterns like those of shallow-water cephalopods are not completely absent, as was revealed to scientists by the Octopus Squid (*Octopoteuthis deletron*). In the absence of a well-lit environment, it had been assumed that all their visual communication with each other would depend on bioluminescence. However, in 2008, scientists were able to use ROV observations to complete an ethogram, or inventory of behaviors, for this species. They found it to be as rich as that of a shallow-water cephalopod. Seventy-six Octopus Squid observed over 15 years revealed distinct skin patterns, body postures, and, certainly, usage of bioluminescent capabilities. Some were comparable to patterns already described from other cephalopods, while others were brand new. What purpose these patterns may serve, or whether they are evolutionary relics, remains to be explored.

Camouflage methods used at different depths

Even in shallow water, cephalopods have adapted to a huge diversity of lighting. Some camouflage with any background in the bright daylight, while some hide during the day and come out at night. Deep-water cephalopods are similarly diverse, with some species that blend into the blackness and others that light it up on purpose. This chart offers a simplified summary of common techniques.

OCEAN ZONE	COLOR/CAMOUFLAGE METHOD
Epipelagic	Color changing, counterillumination
Mesopelagic	Counterillumination, transparency, red
Bathypelagic	Red or black (ink)
Deep benthic	White

Reproduction and growth in the deep

The Octopus Squid (*Octopoteuthis deletron*) is named for having lost its tentacles over evolutionary time. Members of this species also sometimes lose arms deliberately over the course of their lifetime. Like a lizard dropping its tail to distract a predator, but a bit more aggressive, these animals will actually hook their arms onto a predator and then self-amputate to escape.

SEX UNDER PRESSURE

Octopus Squid have achieved some degree of cephalopod fame for their same-sex mating behavior, documented in 2012. Unlike many other cephalopods, male Octopus Squid do not have a hectocotylus or modified reproductive arm, but instead possess a "terminal organ," which is a kind of phallus or penis, long enough to pass spermatophores to another member of their species. Spermatophores are complex packages of sperm that can ejaculate, or discharge, independently upon exposure to seawater, and ideally in close proximity to a recipient. When this happens, a spermatophore rapidly inverts and implants itself into the skin, forming a sac called a spermatangium. These spermatangia are handy for scientists, serving as readily visible markers that a cephalopod experienced a mating event with another cephalopod.

Videos taken by ROVs in the Monterey Submarine Canyon revealed that both male and female Octopus Squid often possess spermatangia. Scientists at MBARI concluded that members of this deep-sea species so rarely encounter another member of their own species that they might as well always try to copulate, even if it will only sometimes produce the evolutionarily advantageous result of offspring.

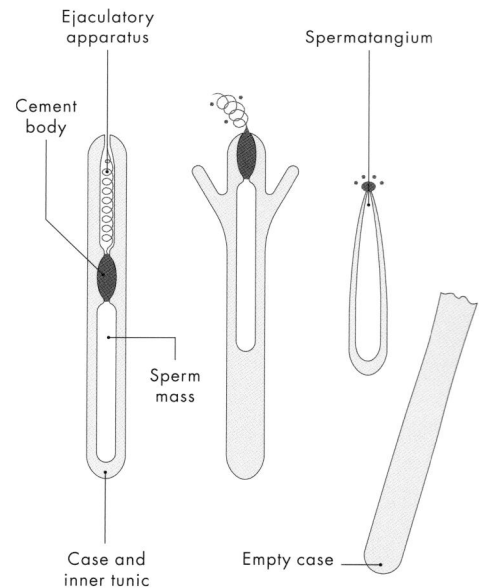

Spermatophoric reaction

The spermatophores produced by male cephalopods are complex structures. During the spermatophoric reaction, a spermatophore turns inside out to create a spermatangium, attached to the female cephalopod with a combination of adhesive cement and sharp stellate particles.

MINIMAL MATING AGGREGATIONS

In 2017, the ROV *Deep Discoverer* recorded putative mating behavior in the deep-sea squid *Chiroteuthis* sp. This was exciting because the genus is known primarily from paralarvae and juveniles, and adults have rarely been observed. On this expedition, three *Chiroteuthis* adults, all sexually mature, were recorded in close proximity. It would be a paltry number of Market Squid (*Doryteuthis opalescens*) or Humboldt Squid (*Dosidicus gigas*)—species that tend to gather in much larger groups—but for *Chiroteuthis* it constituted a significant aggregation, and even led one scientist to suggest that it might represent a mating aggregation.

This hypothesis was supported by evidence that the females had successfully mated, as spermatangia were observed on their bodies. One female, intriguingly, held another squid in her arms that could not be definitively identified but was suggested to be a male *Chiroteuthis*. Since sexual cannibalism is known from other cephalopod species, it's not too much of a stretch to suppose that this female was consuming her mate after acquiring his sperm, and was now acquiring nutrients from his body to help produce her eggs.

EXPLODING EGGS

Octopuses are divided into two groups, cirrate and incirrate. They are named for the cirri, or tiny tendrils of skin, present on the arms of cirrates but not incirrates. More noticeable differences between them include the fins and deep arm webs of cirrates, both useful for swimming in these octopuses' pelagic habitat. The finless incirrate octopuses tend to be benthic, although certain species (such as argonauts) have evolved into open-ocean swimmers. For their part, cirrate octopuses come down to the seafloor to lay eggs. Most do not stay to protect them, but some species lay in protected locations, such as octocorals.

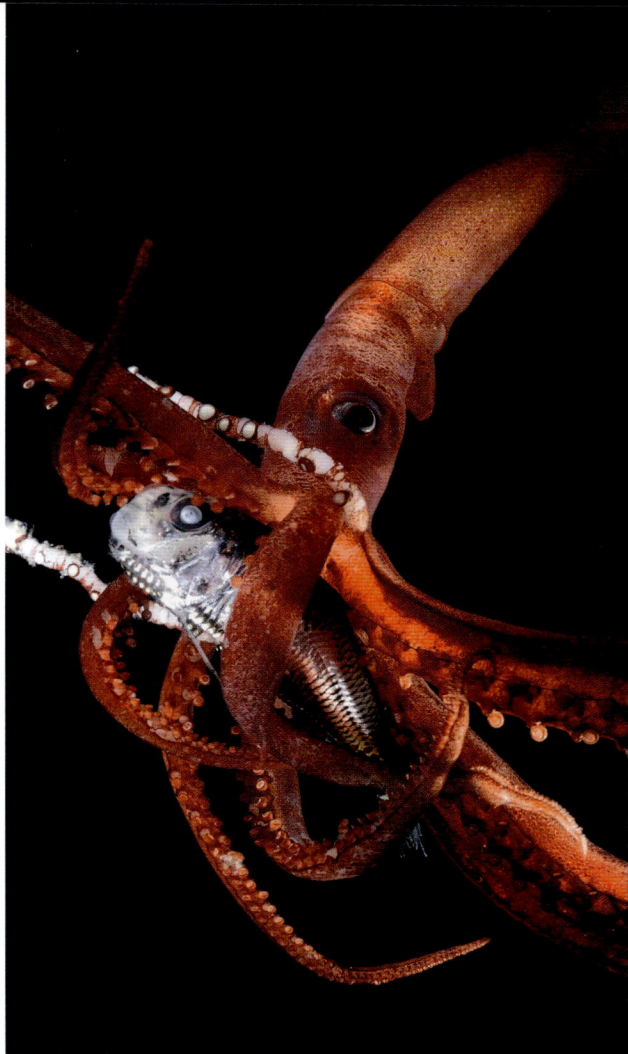

↗   This *Chiroteuthis* consuming a lanternfish demonstrates the species' capacity to consume large prey— including, possibly, their mates.

→   This young Dumbo Octopus from the northwest Pacific is still translucent, its digestive system and gills visible through the skin of its mantle.

Octocorals are deep-sea relatives of the shallow-water reef corals we're more familiar with. Because they live far from sunlight, octocorals don't associate with photosynthetic zooxanthellae. Instead, they are exclusive carnivores, using their tentacles to catch and consume whatever drifts by. Like shallow-water corals, they are colonies of polyps and, also like shallow-water corals, they create an ecosystem by their existence.

Cirrate octopuses take advantage of octocoral structures as a location to attach their eggs. Perhaps because the parents do not remain nearby to offer care, cirrate octopus eggs have an extra protective capsule that incirrate octopus eggs do not have. This thick layer provides additional safety, but also generates a puzzle: how do the baby octopuses get out when they're ready? Bird chicks have an egg tooth to break through their hard shells. Many cephalopods have a hatching gland called a Hoyle organ that produces enzymes to digest the egg material. Cirrate octopuses, however, do not—and yet their eggs are some of the toughest of all cephalopod eggs. In 2017, an ROV shed light on this conundrum with the first observations of cirrate octopus eggs that had, well before hatching, burst through their own tough outer layer. Perhaps the steady growth of the embryo, or osmosis of seawater, produces enough pressure to crack the shell in advance of hatching.

GROWTH RATES IN THE DEEP

A quick race to maturity is typical of most cephalopods, but scientists have wondered whether deep-sea species follow a similarly speedy trajectory. In 2017, researchers published the growth rates and ages of three different deep-sea species: the Swordtail Squid (*Chiroteuthis calyx*), the Cockatoo Squid (*Galiteuthis phyllura*), and the Octopus Squid (*Octopoteuthis deletron*).

Cephalopod statoliths, or inner ear bones, are a crucial tool for scientists studying growth rates. Many, though not all, cephalopods are known to grow their statoliths in daily layers. This can be proven for a particular species by immersing individuals in water that contains tetracycline, a marking chemical that gets incorporated into the statolith, then keeping the animals alive for a certain number of days afterward. If they are then sacrificed so their statoliths can be removed, scientists can count the layers deposited after the tetracycline mark and see if they match the number of days the animal was kept alive—like cutting down a tree to count its rings.

This technique demonstrated that both Swordtail Squid and Octopus Squid produce daily statolith increments. Cockatoo Squid exhibited a less straightforward statolith growth pattern. However, by combining statolith data with other measurements of maturity and size from a number of wild-caught specimens, scientists were able to determine that all three species mature at a greater age than coastal cephalopod species, probably more than a year, and go on to live at least two years, with brooding females likely to have the longest life spans.

↗ Octocorals like this one in the Gulf of Mexico can be habitats for many other species, such as the two squat lobsters seen here.

→ This closeup of a Cockatoo Squid displays the animal's habit of holding all its arms straight up, so that they resemble the crest of a cockatoo. Under-eye photophores block shadows that might give away its presence to predators below.

Dumbo Octopus

Swimming umbrella

SCIENTIFIC NAME	*Grimpoteuthis discoveryi*
FAMILY	Opisthoteuthidae
MANTLE LENGTH	2¼ in (5.8 cm)
TOTAL LENGTH	8¼ in (21 cm)
NOTABLE ANATOMY	Large fins
MEMORABLE BEHAVIOR	Fin-flapping locomotion

Inhabiting the northern Atlantic Ocean, this is a classic cirrate octopus. Species in the family Opisthoteuthidae are sometimes referred to as umbrella octopuses because of their full webbing, with membranes between all of their arms. Such extensive membranes are not seen in typical incirrate octopuses, and highlight the different adaptive purposes of the arm crown in these two different groups. Incirrate octopuses use their arms as sensory, crawling, food-catching appendages. They can sometimes expand short webs between their arms to "umbrella-capture" prey, but having full webs down to the arm tips would inhibit exploratory behavior. Cirrate octopuses, on the other hand, have no rocks or kelp or coral to explore. Their arm crown is useful as a parachute, spread out to prevent them from sinking, or even pulsed like the bell of a jellyfish to help them swim. And it, too, can be used to encompass prey.

Members of the genus *Grimpoteuthis* are also known as dumbo octopuses, due to the large fins that resemble the flapping ears of the iconic baby elephant. These fins are common to all cirrate octopuses, and like Dumbo's ears they are effective for locomotion. That's partly because they have an internal skeleton to work against. Cirrate octopuses have the most substantial internal shell remnant of any octopus: a structure often shaped like a U or V that serves as an anchoring point for fin muscles. Although *Grimpoteuthis discoveryi* engages in active swimming with its fins, it has also been observed settling on the soft seafloor, where this species lays its eggs.

It was a *Grimpoteuthis*, unidentified to species, that recently set the depth record for octopuses in the Indian Ocean. It may belong to one of the existing species, or it may even be a new species. Researchers have pointed out that animals assigned to *Grimpoteuthis* need further study to tease apart which anatomical differences are due to sex or age, and which are truly species-specific traits.

The genus name pays tribute to Georg von Grimpe (1889–1936), a German cephalopod biologist, and the species name is in honor of the ship RRS *Discovery* (RRS stands for Royal Research Ship). The current *Discovery* is a traveling laboratory and scientific research ship, the latest in a long line of exploratory vessels of that name, going back to the Antarctic expedition of 1901–1904 led by Robert Falcon Scott.

Cirrate

Incirrate

Cirrate vs incirrate octopuses

In addition to the cirri (short tendrils lining the arms) that give them their name, cirrate octopuses also have paired fins. Cirrate octopuses also tend to have larger arm webs than incirrate octopuses.

→ Although the fins of a Dumbo Octopus resemble ears, they are attached to the mantle, not to the head, and serve as organs of locomotion.

VULCANOCTOPUS HYDROTHERMALIS

Deep-sea Vent Octopus

Volcano prowler

SCIENTIFIC NAME	*Vulcanoctopus hydrothermalis*
FAMILY	Enteroctopodidae
MANTLE LENGTH	2¼ in (5.6 cm)
TOTAL LENGTH	8½ in (22 cm)
NOTABLE ANATOMY	Bishop-shaped mantle
MEMORABLE BEHAVIOR	Temperature tolerance

Deep-sea Vent Octopuses are stunning examples of the adaptability of octopus anatomy and physiology. Other than their ghostly white color, they look like any octopus you might see on a reef or in a tide pool. Even the striking shape of their mantle, resembling the head of a bishop (if you're not used to looking at real-life bishops, consider the chess piece), is a shape that other octopuses can also create by adjusting their pliable mantles. Deep-sea Vent Octopuses have long thin arms, with webs stretched between them at the bases, and males possess a hectocotylus. There's some indication that they segregate by sex, since all the individuals found at one site were males, with females eventually discovered at a different site.

Their habitat is astonishing. They live thousands of feet deep, in close association with hydrothermal vents where water temperatures can climb well above boiling (the water does not boil because of the tremendous pressure). These habitats are covered with giant tube worms that have no guts but are instead filled with symbiotic bacteria that use the chemicals of the vent to produce food in the same way that plants use sunlight to produce food. The tube worms are almost like the trees of a forest, serving as ecosystem engineers, and crabs and octopuses and other animals crawl on and around them, hiding among the large "trunks," feeding on each other.

Deep-sea Vent Octopuses are thought to eat crabs and shrimps, some of which feed directly on the bacterial mats that grow around the vents. In 2005, scientists observed a "feeding frenzy" of a dozen Deep-sea Vent Octopuses eating amphipods (little shrimplike crustaceans). These amphipods are too small to target and capture one at a time, but they gather in large groups that are worth the octopuses' while. Individual Deep-sea Vent Octopuses seemed to be competing with each other for access to this prey.

A closer look reveals many ways in which their anatomy is indeed quite different from that of a shallow-water octopus. They have no ink sac and no chromatophores. Although they have large eyes, it's not clear if they can see, since they didn't respond to blindingly bright ROV lights.

Deep-sea Vent Octopuses may have adaptations to high temperatures that have not yet been studied, although it would also be quite possible for them to avoid hot water and still benefit from the abundance of food and life in this system.

→ Deep-Sea Vent Octopuses are fairly small and so far known only from vents in the East Pacific. It is possible that they will turn up in other locations as more deep-sea environments are explored in the future.

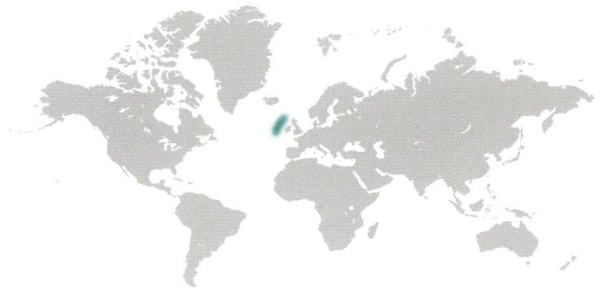

Benthoctopus johnsonianus

Seafloor hunter

SCIENTIFIC NAME	*Benthoctopus johnsonianus*
FAMILY	Octopodidae
MANTLE LENGTH	4¼ in (11 cm)
TOTAL LENGTH	20 in (51 cm)
NOTABLE ANATOMY	Sturdy body
MEMORABLE BEHAVIOR	Relatively deep habitat for the genus

Officially described for the first time in 2006, this deep-sea octopus helped sort out a century-old challenge. Scientists have been puzzling over the genus *Benthoctopus* as far back as the remarkable self-taught marine biologist Annie Massy (1868–1931), who described *B. normani* in 1907. Unfortunately, although Massy's drawings survive, the actual specimen she studied does not. It must have been lost at the museum where it was stored sometime in the intervening hundred years.

Additional specimens gathered over several decades and analyzed in the early 2000s allowed scientists to confirm the identity of Massy's *Benthoctopus normani* and erect a brand new species of *Benthoctopus*, named for the husband of the paper's senior author. *Benthoctopus johnsonianus* inhabits the bathypelagic northeast Atlantic, particularly in the Porcupine Seabight off the coast of Ireland (Massy's home country). It lives deeper than its cousin *B. normani*, and is also distinct in a variety of physical features: its suckers are closer together and its body is stockier with shorter arms and wider head and mantle.

Species in the same genus can often be told apart by their skin structure and pattern, as we've encountered before— for example, the eyelash papillae distinguishing Pacific Red Octopuses (*Octopus rubescens*, pages 104–105) from juvenile Giant Pacific Octopuses (*Enteroctopus dofleini*). However, Benthoctopus species have no such distinctive skin features.

Instead, their definitive identification must be genetic. Also in the Porcupine Seabight in the early 2000s, scientists were excited to observe a *B. johnsonianus* at a large food fall experiment, demonstrating the complexity of the ecosystem such a fall can create (see page 217). Two dead porpoises (they had been killed by bottlenose dolphins) were dropped to the seafloor at bathyal depths and monitored by video. Scavengers, including several different kinds of fish and a whole lot of squat lobsters, came to eat the meat.

At the deeper fall, 8,891 ft (2,710 m), an individual *B. johnsonianus* showed up not to scavenge the porpoise but to feast on the abundance of lobsters. Video recorded a decrease in lobster numbers from 30 to 7 "coincident with the appearance" of the octopus. That doesn't mean that the octopus ate 23 lobsters, but it does indicate that the little crustaceans scattered at the appearance of the large predator. The octopus left and returned several times over the weeks, having presumably identified the porpoise fall as a good place to get a lobster meal.

→ It's possible that further exploration will reveal a broader distribution for *Benthoctopus johnsonianus*. If they have the ability to disperse as either paralavae or adults, they could take advantage of whale falls and other concentrations of food over a large area.

Blind Cirrate Octopus

Sightless swimmer

SCIENTIFIC NAME	*Cirrothauma murrayi*
FAMILY	Cirroteuthidae
MANTLE LENGTH	Estimated 10 in (25 cm)
TOTAL LENGTH	3¼ ft (1 m)
NOTABLE ANATOMY	Reduced eyes
MEMORABLE BEHAVIOR	None—yet

Of all cephalopods, the Blind Cirrate Octopus has the most reduced eyes. They are still capable of sensing light, but not of forming images, since they lack lenses and are embedded within the head, fully covered by gelatinous material. They do have a reasonably well-developed retina and are connected to an existent, albeit tiny, optic lobe. (The optic lobes of shallow-water cephalopods often comprise most of their brain matter.)

The arms and web constitute a substantial portion of the body, making Blind Cirrate Octopuses look at least as much like umbrellas as the cephalopods actually called umbrella octopuses (Opisthoteuthidae). Their fins are quite large, and attached to the center of the mantle rather than the tip. Their suckers are tiny and grow on the tips of long stalks. The name of the genus as well as the family to which they belong refers to the cirri, or feathery projections, that line each arm, even though such cirri are shared by the entire group of cirrate octopuses as well as by Vampire Squid (*Vampyroteuthis infernalis*). Cirri are not modified suckers, but rather flaps of skin that grow outside the rows of suckers. They are thought to be sensory organs that are receptive to touch and therefore able to gather information about contact with minute plankton or even water movement that might indicate a swimming prey item.

These octopuses lack a radula, but they do have a "fleshy tongue-like structure" as described by scientists. They likely feed on small plankton. They can swim by flapping their large fins, like other cirrate octopuses. They have also been observed swimming like a jellyfish, arms and arm web expanded and pulsing. They have a funnel, but are not known to employ it in jet propulsion. Scientists have speculated that jets of water from the funnel, which is quite long and thin, could be directed at the soft seafloor. This could serve to dig up food, or to make a space to lay eggs.

Blind Cirrate Octopuses have been found around the world in bathypelagic regions, with the single exception of a specimen found in an ice hole in the Arctic Ocean right at the surface. They are nowhere abundant, with only five specimens having been found between the species' first discovery in 1910 until 1978, when an expedition encountered three additional individuals.

→ Inside the translucent mantle of this octopus it is possible to spot internal organs: a dark orange gut and whitish gills. The orange color of the arms and web would look black at depth.

ROSSIA PACIFICA

Stubby Squid

Cute crawler

SCIENTIFIC NAME	*Rossia pacifica*
FAMILY	Sepiolidae
MANTLE LENGTH	2 in (5 cm)
TOTAL LENGTH	4¼ in (11 cm)
NOTABLE ANATOMY	Large eyes
MEMORABLE BEHAVIOR	Being adorable

The Stubby Squid gained some degree of notoriety as the "Googly-eyed Squid" in 2016, when ROV video of a Stubby Squid on the deep seafloor off the California coast went viral. The scientists' reaction was recorded live, and we all related to how adorable they found this small cephalopod.

It has a fairly remarkable range for a benthic animal, existing happily on sandy substrates from a shallow 66 ft (20 m) to well over 4,000 ft (1,200 m). There are even reports of Stubby Squid venturing into intertidal regions at night, when the species is most active. Like other sepiolids, it buries itself in sand, first squirting a jet of water to make a concave seat and then using its arms to scoop sand up over its body. Stubby Squid have no photophores, which is consistent with their benthic habits—they don't spend enough time up in the water to benefit from counter-illumination, hiding instead in the sand or mud on the seafloor when they need protection.

They breed in deep water and attach their eggs to solid substrate. They are not exactly semelparous or iteroparous; they only live for one or two years and reproduce at the end of it, which seems semelparous, but females can deposit multiple batches of eggs separated by several weeks, which is more iteroparous. One female can produce several hundred eggs that are substantial in size, larger than those of shallow-water sepiolids, which may be due to the relative lack of available nutrition in the deep sea. Providing babies with extra yolk so they can grow larger before hatching gives them a head start on finding their own food.

The eggs also have the added protection of hard shells, something unusual among cephalopods and among marine animals generally. These seem to deter predators for the lengthy developmental time of nine months. Yes, these tiny bobtail squid spend as long in the egg as you spent in your mother's womb, even though they will only live for a year or two after hatching! Their development is likely slowed by the cold water in the deep sea. Stubby Squid hatchlings are not paralarvae, but benthic juveniles that promptly take up the habits and habitats of their parents. They eat shrimp, primarily, but will also take fish and mollusks.

→ Although in many places the Stubby Squid only gained notoriety because of a cute video, around Japan this species is caught by trawl and regularly eaten.

Seven-arm Octopus

Giant mama

SCIENTIFIC NAME	*Haliphron atlanticus*
FAMILY	Alloposidae
MANTLE LENGTH	27 in (69 cm) (female)
TOTAL LENGTH	13 ft (4 m) (female)
NOTABLE ANATOMY	Female size
MEMORABLE BEHAVIOR	Lurking near the seafloor

Seven-arm Octopuses belong to the superfamily Argonautoidea, which also includes argonauts, blanket octopuses, and tuberculate pelagic octopuses. They derive their common name from the males of the species, which are smaller than females and possess an enlarged detachable hectocotylus, like argonaut males. This modified arm is completely embedded in a sac below the male's right eye until the time comes for it be released. As the most gelatinous of all argonaut relatives, this species' hectocotylus is difficult to spot, so males appear to have only seven arms. In 1929, a male *Haliphron atlanticus* was even described as its own species, *Heptapus danae*.

However, as in the other members of the Argonautoidea, it's the females that are really distinctive. In the case of *Haliphron*, they are absolutely gigantic, far greater in total length than the height of a human. Even the males are not true dwarf males like those of other argonautoids, since they reach up to 4 in (10 cm) in mantle length and twice that in total length. Seven-arm Octopuses are thought to spend much of their lives close to the deep seafloor. They also have been found in shallow water, though, especially the males, which have been documented interacting with jellyfish like other members of Argonautoidea. Also like these relatives, females are believed to brood their eggs. Hatchlings, unlike those of many other pelagic species, are not found in shallow water and are thought to occupy deep habitat like their mothers.

The funnel of Seven-arm Octopuses is completely embedded in and continuous with the skin of the head, rather than being a separate organ below the head as is typical of cephalopods. Juvenile females appear to be fairly muscular swimmers, but they become more and more gelatinous as they age. When large females are removed from the water, they spread out like a jellyfish or a blobfish, unable to keep their shape. Like many other deep-sea animals, the structural integrity of their body depends on the pressure of the surrounding water.

As one of the largest octopus species, the Seven-arm Octopus may be responsible for some myths and legends of the Kraken, especially when specimens are badly damaged and open to interpretation, as occurs when they are caught in a net or found floating dead on the surface. Fish, mammals, and birds all eat them.

→ Count the arms of this paralarval Seven-arm Octopus, and you'll find eight. This marks it as a female of the species. If she survives these tender early days, she will grow to be one of the most substantial octopus adults in the world.

Deep-sea Squid
Muscle of the deep

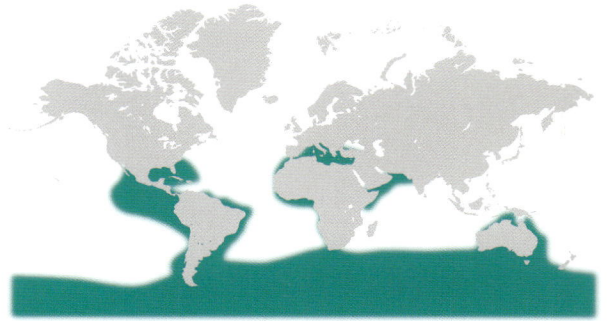

SCIENTIFIC NAME	*Bathyteuthis* spp.
FAMILY	Bathyteuthidae
MANTLE LENGTH	3¼ in (8 cm)
TOTAL LENGTH	8¼ in (21 cm)
NOTABLE ANATOMY	Extraocular photoreceptors
MEMORABLE BEHAVIOR	Brooding of egg sheets

Obviously, this is not the only deep-sea squid, so instead of using an ambiguous common name, we'll stick with the scientific name, which encompasses three species. They are close relatives of *Chtenopteryx* spp., the comb-finned squids, which have those strange stiff ribs lining their fins. *Bathyteuthis* grow more typical squid fins, small and round at the tip of their mantle.

Bathyteuthis eyes point "forward," which for a squid means toward the arms, which also means toward any prey they would be likely to grab. They have photophores at the base of their arms, the purpose of which is unknown. Just behind their eyes, *Bathyteuthis* have "extraocular photoreceptors" or bits of anatomy that are not eyes but are able to perceive light. They are not image-forming, but are tuned to pick up bioluminescence that the eyes might not be able to register.

→ Many midwater and deep-sea squid use lightweight ammonium to achieve the neutral buoyancy their ancestors lost when giving up their shells. *Bathyteuthis*, by contrast, stores some other as-yet-unidentified lightweight material, which allows it to build more muscle and therefore swim more strongly than ammoniacal squid.

Scientists speculate that they might even register the bioluminescence of tiny plankton that get accidentally sucked into the mantle cavity during breathing. This would alert the squid to the fact that they're unintentionally glowing from the inside, which might attract predators, and they can cover up with chromatophores or breathe out the troublesome plankton guests.

Bathyteuthis squid are found all around the world. The first species described was *B. abyssicola*, which is abundant in the Southern Ocean, where it camouflages itself with deep-sea red coloration. This species has also been found in the Pacific, Atlantic, and Indian Oceans, and undergoes diel migration in all these locations. Another species, *B. bacidifera*, shares space with *B. abyssicola* in the eastern tropical Pacific and Indian Oceans, but can be distinguished anatomically by wiggly trabeculae, similar to papillae, at the base of its arms.

Finally, there is *B. berryi*, also found in the eastern Pacific but further north in temperate waters. They do not have the trabeculae of *B. bacidifera*, and instead their arms bear an incredibly dense number of tiny suckers. In 2012, ROV observations revealed that this species holds its eggs in its arms and broods them until hatching. The eggs are embedded in a structure that is not quite a mass, nor a capsule; scientists call it an egg sheet. Rows of white eggs are embedded in a flat sheet of mucus, longer than the body of the brooding animal. It's believed that all *Bathyteuthis* are brooders, although the other species have yet to be observed engaging in this behavior.

ANTARCTICA
& THE ARCTIC

Introduction to the poles

The north and south polar regions are the only places on Earth that point completely away from the sun for part of the year, and completely toward it for another part of the year. This gives rise to the midnight sun in summer, and the less-poetic midday night in winter. Throughout our planet's life span, this is the only feature of the poles that has remained constant.

A BRIEF POLAR HISTORY

Today we associate both polar regions with ice and cold, but that hasn't always been the case. At warmer times in Earth's history, there have been no polar ice caps, and forests have thrived at extreme latitudes. We're now in a planetary cold spell, which started 3 million years ago—long enough that species have adapted specifically to this climate. Within this larger ice age, glacial periods come and go over tens of thousands of years. We're between glacial periods right now, with human-caused climate change altering our future trajectory for as much as the next half a million years or even more (see pages 264–265). Sea ice at the poles forms a critical component of our planet's climate, because its white surface reflects sunlight that would otherwise be absorbed. When sea ice retreats, this sunlight warms the seas, just like it would warm you more if you wore a dark blue outfit instead of a bright white one.

A UNIQUE ENVIRONMENT

Sea ice grows and shrinks at the poles over the course of the year, expanding its coverage in the winter and retracting it in the summer. Terrestrial and air-breathing animals are sensitive to these shifts. Emperor penguins in Antarctica walk to the thickest part of the sea ice to lay and incubate their eggs, so the eggs won't risk

↗ Penguins are typically associated with the South Pole. Penguins are indeed limited to the Southern Hemisphere, but some species live in ice-free, even tropical, latitudes.

← Some Arctic ice melts and breaks up every summer, then freezes anew in winter, but climate change has tipped the overall balance toward ice loss.

falling into the water. Polar bears in the Arctic can swim, but not indefinitely, and they depend on solid ice not only to raise their young but to rest and sleep. Ice strong enough to walk on is good for some animals, while ice thin enough to break is critical for marine air-breathers and animals that fish for food through ice holes.

The long days and nights of the polar winter and summer mean that for much of the year there's no stimulus for diel vertical migration. Indeed, researchers have found that the Arctic deep scattering layer, while it exists, does not migrate during winter and summer. Additionally, at both poles animals have evolved incredible adaptations to the cold. Air temperatures can drop as low as the world record of −129 °F (−89 °C) in Antarctica. Sea temperatures remain comparatively balmy, because as long as you're in liquid water, it's not going to drop below freezing. Or is it?

SALT AND ICE

The process of freezing shares something with the process of evaporation. When water shifts phase from liquid to either gas or solid, it can't carry salt along. Similar to how evaporation increases the ocean's salinity, so does the formation of sea ice. As below-freezing air in contact with surface seawater causes it to freeze, the salt is excluded from the ice, remaining dissolved in the water below.

The freezing point of salt water is lower than that of fresh water. This is why transportation departments use salt to de-ice roads. As long as there's even a little liquid water available (which de-icing trucks sometimes add in the form of brine), the salt can mix with it and lower its freezing point, and any additional ice melt will also take up salt and be that much less likely to re-freeze.

Let's take a moment to understand how salinity is measured. Typical seawater salinity is 35 parts per thousand, which means 35 g (7 teaspoons) of salt per 1 kg of seawater. Parts of the sea close to estuaries or with lots of freshwater runoff can have much lower salinities,

all the wa down to zero parts per thousand in the middle of a freshwater plume, which no cephalopod we know of can tolerate. The species that are adapted to life at lower salinities (such as the Brief Squid, *Lolliguncula brevis*, and the Giant Cuttlefish, *Sepia apama*) can handle salinities just barely below 20 parts per thousand.

Another influence that can reduce ocean salinity is ice melt. When seasonal ice that formed in the winter melts back into the sea in the summer, or if Antarctica's or Greenland's or other glaciers melt into the sea, this reduces salinity at the sea surface. Because it has less salt, it is less dense and tends to float. Some scientists theorize there was a time in our planet's history when the Arctic Ocean was isolated from other oceans by land masses, and acquired a significant layer of freshwater on top of the salt water from high rainfall and river runoff. About 49 million years ago, this layer may have supported the growth of enormous quantities of a freshwater plant called *Azolla*, and this bloom trapped enough carbon dioxide when it sank to significantly change Earth's climate.

A GLOBAL CIRCULATORY SYSTEM

The saltier water is, the colder it can get before freezing, so the salty water left behind when sea ice forms is the coldest liquid water in the entire ocean. Both low temperature and high salinity contribute to increasing its density, so it sinks. This cold, heavy brine spreads over the bottom of the polar basins, then flows into the Atlantic and Pacific basins, following the topography of the seafloor and driving the ocean conveyor belt.

We've already met this cold water, but now we know where it comes from—sea ice forming at the poles. This is sometimes referred to as thermohaline ("thermos" for temperature and "haline" for salt) circulation. Amazingly, it forms currents so stable they can traverse the globe without petering out or changing direction.

The cold dense water from the Arctic flows south, from Greenland through the Atlantic all the way to Antarctica, where it is recharged with brine at this opposite pole. Circling the continent, it then flows north in two currents, into the Pacific and Indian Oceans. Here it gradually warms and rises, creating nutrient-rich upwelling. Near the surface in the northern parts of these oceans it loops back around to go south again, mixing with other surface water, crossing the Indo-Pacific into the South Atlantic and eventually arriving in the Arctic to be chilled again.

↙ This Alaskan ice flows at a "glacial" pace from the land into the sea. Glaciers that reach the water can melt at a surprising rate, as ocean waters warm due to climate change.

↓ A visualization of water movement in the North Atlantic shows the Gulf Stream current that flows from the Gulf of Mexico along the coast and eastward toward Europe. Deeper currents below carry water in the opposite direction.

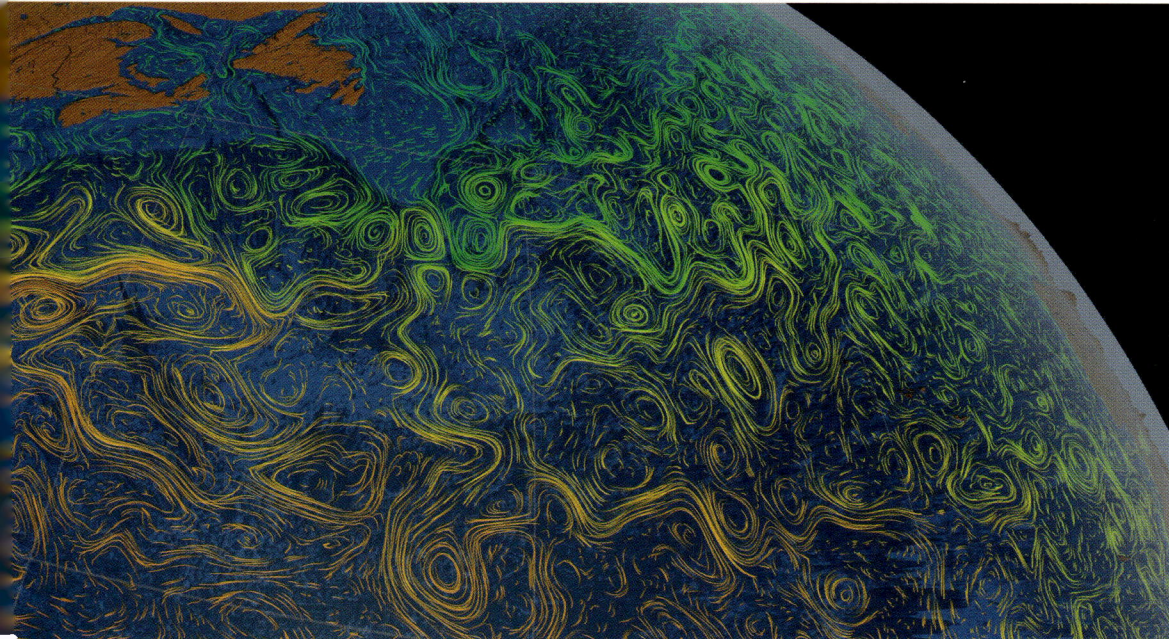

The Arctic

While the Antarctic is a land mass surrounded by ocean, the Arctic is an ocean surrounded by land. For the length of our existence as a human species, the Arctic has been covered by a cap of perennial ice—ice that never melts, even in the summer. This is why early explorers were able to travel by sled and ski and foot to the pole, and had somewhere to plant a flag even though they stood atop an ocean.

LOSS OF SEA ICE

However, in the later years of the twentieth century and the beginning of the twenty-first, sea ice in the Arctic has been declining. Some of it has always melted in the summer and regrown in the winter, but lately the melting has been more extensive and the summer season has lasted longer, while the time of ice formation in the winter has gotten shorter. The Northwest Passage between the Atlantic and Pacific Oceans over the top of Canada has become viable for transport due to the loss of sea ice, leading to debates over shipping rights. As Arctic ice retreats, additional resources are likely to become available, including oil and fishing grounds. Human reactions to the changes range from excitement about new opportunities through to worry about the potential for conflict and concern for negative impacts of human exploitation on the Arctic ecosystem— which is already suffering directly from climate change.

GEOGRAPHY

The Arctic Ocean is surrounded by various land masses (Northern Europe, Greenland, and North America) and smaller seas. Greenland, the land closest to the North Pole, houses one of our planet's two ice sheets—an expanse of ice that covers more than 20,000 square miles of land. When glaciers or ice sheets extend from land onto the ocean, they're called ice shelves. Canada has ice shelves that go into the Arctic Ocean,

September

March

Seasonal ice at the Arctic

As the North Pole tilts away from the sun, temperatures drop and seawater freezes. It continues to freeze over the course of the winter, resulting in the greatest extent of Arctic sea ice usually in March. The reverse happens in the summer, with the smallest extent of sea ice occurring around September.

though three have broken up: one in the 1960s, one in 2005, and one in 2008. One of the remaining four, the Milne Ice Shelf, began collapsing in 2020.

Between the Arctic Ocean and northern Europe (Russia to Norway) lies the Barents Sea, a large expanse of polar water filled with a diversity of animals. Looking down on the North Pole and traveling clockwise from the Barents Sea, one bumps into Greenland, which blocks oceanic passage down to the Atlantic. On the other side, between Greenland and Canada, is Baffin Bay, which is also rich in marine life.

↑ An island in Northwest Greenland is covered with a blanket of snow and ice. Sea ice along its coast floats on the water.

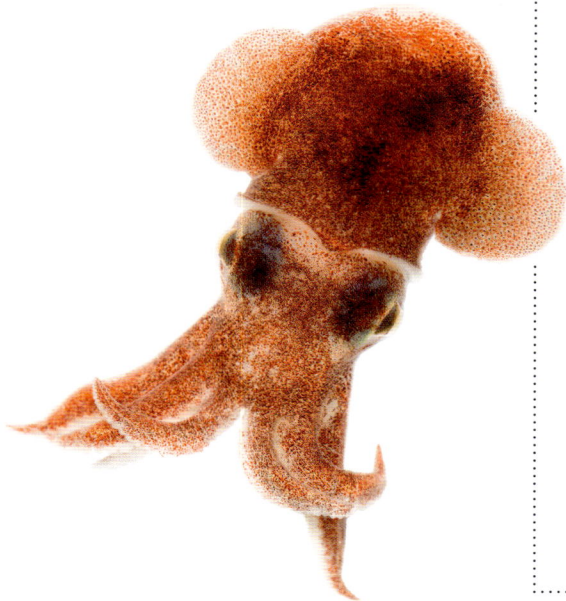

↘ *Rossia palpebrosa* is a bobtail squid that lives in a range of habitats and depths in the Arctic and the northern Atlantic.

ARCTIC CEPHALOPODS

Arctic waters tend to have low salinity, due to the frequent ice melt, which is not generally great for cephalopods. The two cephalopods most often found here are the Warty Bobtail Squid (*Rossia palpebrosa*, pages 274–275) and the Boreo-atlantic Armhook Squid (*Gonatus fabricii*). The latter is the only Arctic pelagic cephalopod. Estimates of biomass suggest that there are well over a billion individual Boreo-atlantic Armhook Squid in the Barents Sea alone. Although this is a whole lot of squid, it's well under the abundance of other invertebrates and fish in the area, suggesting that Arctic ecosystems are not as cephalopod-heavy as Antarctic ones. However, the Boreo-atlantic Armhook Squid is still a common prey item for fish, birds, and mammals, and although humans are not currently catching it for food, they're considering doing so. In 2022, an expedition found one of these squid smack in the middle of the Arctic Ocean, where they had only expected to find much smaller planktonic life, confirming that despite the "Boreo-atlantic" limitation in its name, it is a truly Arctic species.

Antarctica

Antarctica is a major land mass, a continent that experiences plate tectonics like all the other continents. Once part of a mega-continent that included South America, Africa, and Australia, Antarctica lay further north and supported a warm, lush ecosystem. As the supercontinent broke apart and Antarctica moved toward the South Pole, its distance from the other continents opened up space for ocean currents to flow continuously around Antarctica, isolating its climate and driving further cooling.

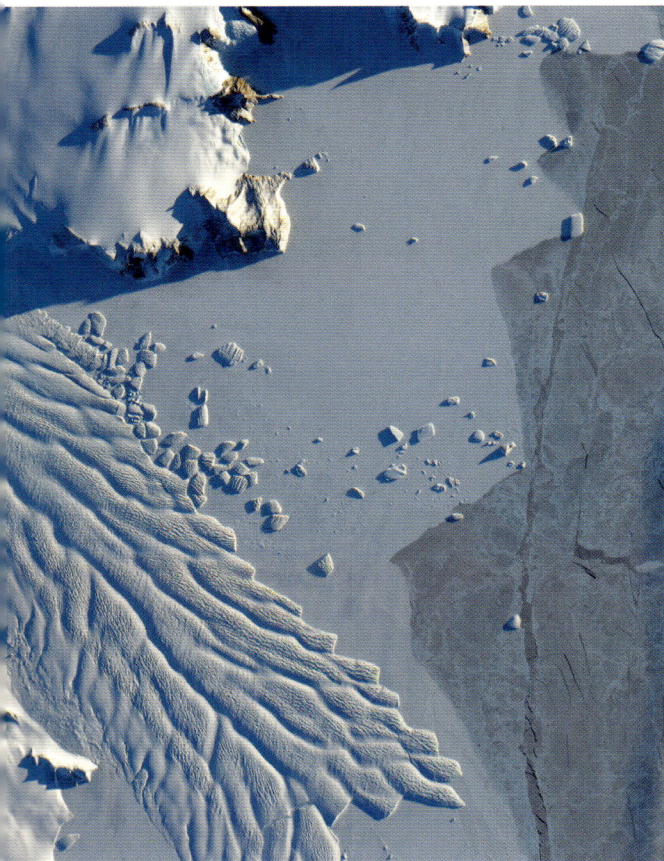

SOUTHERN ICE

Antarctica is now a combination of land mass, ice sheet, and ice shelves. Under, around, and beyond its ice shelves lies the Southern Ocean, home to incredible biological diversity and fascinating evolutionary trajectories. Here is where the heaviest squid in the world, the Colossal Squid (*Mesonychoteuthis hamiltoni*) is found, a possible example of polar gigantism (pages 196–199). The solid land of Antarctica is covered by an ice sheet, as well as housing numerous glaciers. The ice sheet is one of only two on the planet and it is the largest, covering 5.4 million sq miles (14 million sq km). Its exclusive input is snowfall, while it loses ice at the edges due to melting or calving icebergs.

One of the strangest properties of water is that its solid form is less dense than its liquid form. Because ice floats on water, ice shelves form where an ice sheet or a glacier reaches off land into the sea. How much of our planet's structure depends on the peculiar features of water! Not all the ice in Antarctica is created by snowfall, though; sea ice can also form from freezing of the ocean water itself.

While winds in the Arctic push warm water underneath ice and increase the likelihood of melt, for a couple of decades winds in the Antarctic swept newly formed sea ice away from the continent, encouraging the

↑   The largest glacier in the world, Lambert Glacier, flows from East Antarctica into the sea.

↖   Krill form a vital part of the Antarctic ecosystem. During the long winter, when food is scarce, they can survive extended periods of starvation and even shrink in size.

←   Like many other glaciers of Antarctica, Matusevich Glacier flows like a very slow river between geological features. Where it reaches the sea, icebergs calve off.

formation of new ice. Thus, unlike the situation in the Arctic, for much of the early twenty-first century Antarctic sea ice covered more area than it had on average before. But when wind patterns changed in 2017, sea ice coverage dropped, and has remained low since.

GEOGRAPHY

Circling Antarctica is the Southern Ocean, which connects to several other oceans at a distinctive natural boundary called the Antarctic Convergence. This is where cold polar water hits warmer water from the Atlantic, Pacific, or Indian Oceans, mixing and pulling up nutrients to support a rich ecosystem. This is where krill are maximally abundant, enough that huge whales migrate enormous distances to feed on them.

There are several seas closer to the continent that also bear mentioning. The Weddell Sea nestles behind the Antarctic Peninsula, a finger of land that reaches up as if to touch the cape of South America. Several ice shelves sit on parts of the Weddell Sea, and it is an important area for research and biological activity. Almost opposite the Weddell Sea is Prydz Bay, Antarctica's third largest bay, covered by the Amery Ice Shelf, which is created by the world's largest glacier, Lambert Glacier, entering the bay. Then, nearest to New Zealand though not very close to it, is the Ross Sea, with the gigantic Ross Ice Shelf.

Adventure in the ice

Our human interest in exploring the poles is a clear manifestation of our desire for making the unknown known. Specific goals have also been set for different expeditions: for fame and glory, for claiming geopolitical clout, to look for resources such as oil or fish, and for knowledge. It is that last search, for knowledge, that drives science, although it is often entangled with other motivators. Sure, it's great to learn new things, but who learns them first? Those who discover new species get to name them, a limited kind of fame. Research vessels and stations also give geopolitical clout. And scientists are heavily employed in both finding various resources and determining the impacts of their extraction.

INTERNATIONAL POLAR YEARS

Beginning at the end of the nineteenth century, researchers from ten European countries plus the United States came together in the first International Polar Year, 1882–1883. They worked from twelve Arctic and two Antarctic research stations to study the weather, atmosphere, currents, and other geological and oceanographic phenomena of the polar regions. (No biology was included.) The second International Polar Year was carried out in 1932–1933, this time with 44 participant nations. Again, the focus was on weather forecasts and mapping for radio communications as well as transportation.

An International Geophysical Year was conducted in 1957–1958, inspired by the International Polar Years and sometimes called the third IPY. The International Polar Year(s) of 2007–2009 is considered the fourth. This was the first time that an IPY explicitly included a biological component. The ongoing focus on geology and meteorology was finally joined by research on the ecology and biodiversity of both terrestrial and marine polar systems. The extensive research and collections made during this time have provided much of the information on cephalopod species in this chapter.

SHIPBOARD SCIENCE AND LABORATORIES

Exploration and research at the poles have always faced the challenge of ice. "Icebreaker" ships have been around since the days of wooden sailing vessels, when metal reinforcement was used to make a hull sturdy enough to break through ice. Eventually icebreaker ships powered by steam, diesel, diesel-electric, and even nuclear power were developed. Icebreakers are used to open passages, escort cargo ships, and for rescue

↑↑ McMurdo research station in Antarctica has all the infrastructure of a small town and is capable of supporting more than a thousand residents.

↑ In addition to permanent stations and large icebreaker ships, scientists sometimes use small vessels to explore and study the polar seas.

↖ This 1909 photograph documents an American expedition that claimed to have reached the North Pole. As the exact point lies below drifting sea ice, it cannot be marked permanently.

missions, but also as research vessels. Icebreaker ships regularly open the way to supply scientific stations in Antarctica, notably McMurdo Station of the United States' National Science Foundation in the Ross Sea.

Although there are no cities or towns at either pole, Antarctica is home to numerous permanent research stations. The countries that operate them have all signed the Antarctic Treaty, an agreement dating back to the beginning of the Cold War in 1959, just after the International Geophysical Year of 1957–1958. The Treaty is an agreement to keep Antarctica set aside for scientific research, prohibiting its use for military purposes.

The Arctic is also home to numerous research stations, most of which are established on land or on ice on top of land, but some are drifting ice stations. Icebreaker ships are crucial for bringing people to and from such stations. As Arctic sea ice loss continues, drifting ice stations are being replaced with large research vessels.

Adaptations to the cold

Although we humans typically exhibit a strong preference for swimming in warm tropical water over cold polar water, animals that actually breathe in water find that polar seas have a distinct advantage over the tropics: plenty of dissolved oxygen. Cold water can hold more oxygen than warm water, and these seas are well mixed by currents and upwelling, unlike the more permanently layered temperate and tropical oceans.

BREATHING IN THE COLD

Antarctic fish can acquire oxygen with little or even no hemoglobin in their blood. Icefish, incredibly, are the only vertebrate ever shown to have no hemoglobin at all. They still have blood, and it picks up dissolved oxygen through diffusion across the animal's gill membranes and thin skin. We don't (yet) know of any cephalopods that have similarly evolved a total loss of hemocyanin, the analogous respiratory pigment in this group.

As it chills, hemocyanin loses its ability to give oxygen to the cells that need it. One way of coping is to ramp up the amount of hemocyanin in your blood,

and at least two species of Antarctic octopus have blood with significantly more hemocyanin than warm-water octopuses. Each Antarctic species also demonstrates further adaptations.

The Giant Antarctic Octopus (*Megaleledone setebos*) has special hemocyanin that doesn't change how tightly it holds oxygen with temperature. Meanwhile, *Pareledone charcoti* produces hemocyanin that can adapt to different temperatures, holding oxygen more loosely at colder temperatures. *Pareledone* octopuses have also been shown to adjust their nerves to better transmit signals in the cold.

POLAR GIGANTISM

The higher availability of oxygen is one explanation that has been proposed for polar gigantism—why certain invertebrate groups evolved extraordinarily large species at the poles. The classic example is a rather unassuming group called sea spiders, which live all over the world but are mostly tiny, transparent, and rarely noticed. In Antarctica, they can be as big as your head.

Not every invertebrate at the poles is a giant, but it's notable that one of the two largest cephalopod species in the entire world, the Colossal Squid (*Mesonychoteuthis hamiltoni*), is a resident of the Southern Ocean.

Another possible explanation for why animals can be large at the poles is, ironically, their slow growth rate, which lowers their metabolism and lengthens their lives. Polar and deep-sea cephalopods tend to develop more slowly and live longer than their warm-water relatives, but most of the slowed development occurs in the egg. A warm-water species might spend only a few days as an embryo, then hatch and live several months as a juvenile and adult. Polar or deep-sea species may spend years as an embryo, and then after hatching live the same number of years as an adult, or even fewer. Why? The mystery of the cephalopod's short life span continues to puzzle scientists.

Cephalopod giants

Extra-large cephalopod species seem more likely to be encountered in cold water. These include the Antarctic-affiliated Colossal Squid and the Giant Pacific Octopus of the chilly northeast Pacific. Even species such as the Humboldt Squid, whose range spans the equator, tend to grow larger at the colder limits of their habitat.

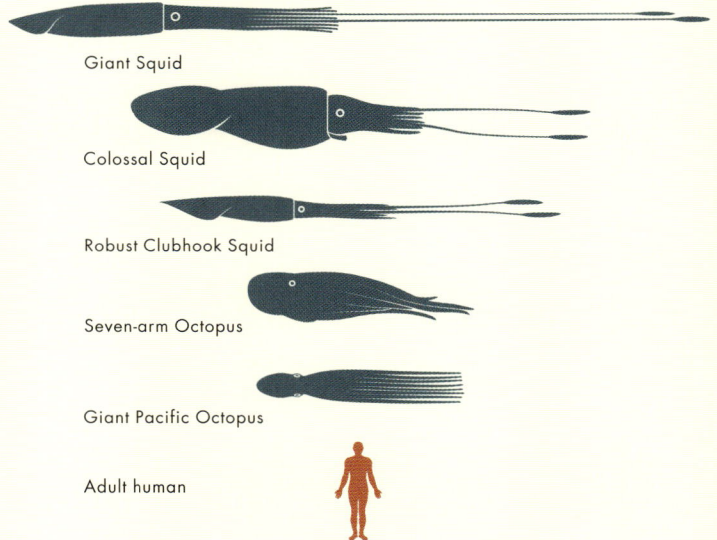

Giant Squid

Colossal Squid

Robust Clubhook Squid

Seven-arm Octopus

Giant Pacific Octopus

Adult human

ANTIFREEZE

Freezing typically kills organisms when the water in their bodies forms ice crystals that tear membranes, causing irreparable damage. Numerous polar species have evolved biological antifreezes, proteins that bind to tiny ice crystals, preventing them from growing large enough to damage membranes. Although no cephalopods are known to have true antifreeze proteins,

evidence suggests that other proteins in their bodies have adapted to the cold. This is the case for the oxygen-binding protein hemocyanin, as we saw, as well as the venom of Antarctic octopuses

Like other octopuses, these species produce protease in their salivary glands to break down the tissues of their prey. These toxins typically become less functional at cold temperatures, to the point of being useless below 39 °F (4 °C)—temperatures at which Antarctic octopuses spend most or all of their lives. Analysis of octopuses collected during 2007–2009 showed that four different species each possessed at least one venom protein that was "extreme cold-adapted" and able to function even below freezing.

← Pygnogonids, or sea spiders, often have eight legs like land spiders, but many species such as this one have ten. Oxygen can be exchanged across their long limbs, so they have no gills.

←← Because unicorn icefish like this one lack hemoglobin in their blood, they are sometimes also referred to as white-blooded fish.

A polar origin
for cephalopods?

Similar in some ways to the Indo-Pacific, the Antarctic is a center of biodiversity and speciation. It is, in a way, a large island, isolated from other continental land masses for the past 35–40 million years or so, which is plenty of time for an endemic fauna to evolve. (Think of all the strange animals unique to the island continent of Australia, which has been similarly isolated for just a few million years longer.) However, the waters around Antarctica are also connected with the global oceans via thermohaline circulation. Many adaptations to the cold water around Antarctica could remain adaptive in the cold deep sea.

OCTOPUS INVASION

The water of the Southern Ocean, moving from Antarctica into the depths of the Pacific and Indian Oceans, could carry polar species with it. Scientists have theorized that polar representatives of many different groups could have invaded and colonized the deep sea, but it's been difficult to test. Eventually, advances in molecular biology and a sufficient quantity of available specimens made it possible to study this possibility for the group of incirrate Antarctic octopuses that includes *Pareledone*, *Megaleledone*, and their relatives.

← A scuba diver equipped for ice diving in the Antarctic has a drysuit and a light to illuminate the astonishing diversity of life in this frigid environment.

↗ Several years after the *Challenger* expedition, the German naturalist Carl Chun led a follow-up called the *Valdivia* expedition, which spent a month exploring and collecting around Antarctica. Chun's careful drawings and reports of cephalopods, published as *Die Cephalopoden* in 1910, greatly advanced the field.

For years even before the International Polar Years, collections of octopuses from both Southern Ocean and deep-sea regions had been underway. In 2008, the genetics of these samples were analyzed to reconstruct the species' evolutionary history and answer the question of whether deep-sea octopuses evolved from Southern Ocean ones and, if so, how long ago.

The group includes six different genera, most of which contain several species. Three of them are deep-sea genera: *Thaumeledone*, *Graneledone*, and *Velodona*, with some species in these genera found near Antarctica, some species found further north but still in the southern hemisphere, and very few species in the northern hemisphere. DNA results indicated that these three deep-sea octopus genera are most closely related to each other, and that this subgroup fits within the set of polar genera, as if it evolved from one of these and then quickly speciated to fill a variety of deep-sea niches. The species among those three deep-sea genera that do live around Antarctica appear to have originated earlier than the ones that live further north in the deep sea, suggesting that they took time to migrate farther from their polar origins.

The timing of their diversification seems to line up with our geological understanding of Antarctica's growing ice sheet and the formation of thermohaline circulation. About 15 million years ago, the East Antarctic Ice Sheet began to expand. The cooling climate and freezing seawater created a greater influx of cold dense water to the deep sea, possibly carrying polar octopuses out into the depths of other oceans. At the same time, active movement by the octopuses may have been encouraged by the fact that Antarctic shallow water was becoming a more challenging environment.

However, Antarctica did not become uninhabitable to octopuses. The remaining three genera in the study, *Adelieledone* (named for the same Adèle as Adélie penguins, she was married to a French explorer of Antarctica), *Pareledone*, and *Megaleledone*, continue to thrive on the shelves and slopes of this continent.

A melting future

In the billions-year-long history of our planet, polar ice caps have come and gone. From this perspective, losing our ice due to climate change may seem like part of a normal sequence of events. However, as with every other aspect of human-caused climate change, this is the first time that ice melt has happened at the current rapid rate. We know that it's because of pollution, and that we have the capacity to slow or even potentially stop the loss.

HISTORY OF ICE CAPS

A hundred million years ago, in the time of ammonites (and dinosaurs), Earth didn't have anything like our current polar ice caps. It's difficult to be completely sure, but scientists have techniques for measuring ancient climate and sea level. Evidence of past sea level changes paints a picture of the periodic formation of small polar ice caps, which locked up water and reduced sea level, then melted to bump sea level back up.

Long after the mass extinction that took away ammonites (and, fine, dinosaurs) around 33 million years ago, Antarctica froze over. This may have been driven by continental movements and possible atmospheric changes. The Arctic didn't join in the ice cap fun until a mere 3 million years ago. This cooling event is thought to have been driven by the formation of the Panama Isthmus between North and South America, which blocked the flow of warm water around the equator and left the poles even more isolated from temperate influences. Thus began our current ice age.

WELCOME TO THE ICE AGE

Although the term "ice age" isn't always used with its exact scientific definition in mind, technically we are currently in one. Within an ice age, extreme cold conditions come and go, and these are referred to as glacial and interglacial periods. We are now in a warmer

Panama Isthmus
North and South America originally belonged to separate supercontinents. As both broke up, the Americas drifted toward each other, and colliding plates pushed up islands that were eventually joined by sedimentary deposits, until the isthmus grew thick enough to permanently separate the Pacific and Atlantic Oceans.

interglacial, which is why we often casually refer to the previous glacial period 20,000 years ago as the "ice age." Based on what geologists have figured out from studying ice cores and deposits, we could expect another glacial period to begin in about 50,000 years—if we hadn't taken the route of global warming. As it is, scientists see no reason to expect another glacial period in the next half a million years, or even further out.

↖ The term "pollution" encompasses a broad range of environmental impacts, from plastic waste and oil spills to the invisible greenhouse gases that cause climate change.

↗ The Lesser Flying Squid lives in several disjunct populations, including the Australian coast, the waters around Mauritius, and the Mediterranean Sea. Its northest Atlantic population has recently extended into the Arctic.

WARMER POLES AND CEPHALOPODS

Climate change has influenced cephalopods in the Arctic already. The Boreo-atlantic Armhook Squid (*Gonatus fabricii*) has expanded its range, and three boreal species have moved in. Research cruises conducted between 2006 and 2011 in the Arctic provided the first records ever of two cephalopod species that had only been known from more temperate waters: the Atlantic Cranch Squid (*Teuthowenia megalop*s) and the Lesser Flying Squid (*Todaropsis eblanae*). Atlantic Cranch Squid live in fairly deep water, and researchers suggest that warm deep-water currents brought it to the Arctic. As for the Lesser Flying Squid, it lives near the continental shelf and slope, and was probably carried by near-bottom currents into the Barents Sea. A third species, the European Flying Squid (*Todarodes sagittatus*), was found in the Arctic in foraging groups in 2010. This species had been seen in the Arctic before, but not for the 25 years prior to this observation.

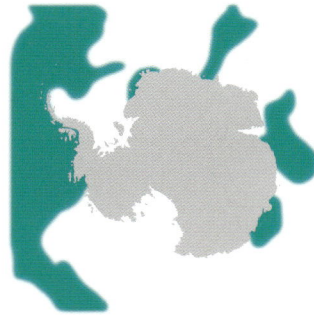

GALITEUTHIS GLACIALIS

Antarctic Glass Squid

Long-tentacled South Pole specialist

SCIENTIFIC NAME	*Galiteuthis glacialis*
FAMILY	Cranchiidae
MANTLE LENGTH	20 in (50 cm)
TOTAL LENGTH	6½ ft (2 m) (unconfirmed)
NOTABLE ANATOMY	Long tentacles
MEMORABLE BEHAVIOR	Abundance around Antarctica

These squid are found throughout the region known as the Antarctic Convergence, and they are the only glass squid other than the Colossal Squid (*Mesonychoteuthis hamiltoni*) that live south of the Antarctic Polar Front. Their diet is mostly krill, and they are eaten by albatrosses and elephant seals, making them a prime example of a nutrient vector—a species that gathers nutrients from the bottom of the food web and makes them available at the top of the food web.

Like other glass squid, Antarctic Glass Squid are not very muscular and have a translucent to transparent body. Studies on their statoliths indicate that they grow more slowly and live longer than tropical glass squid. They have photophores around their eyes.

Antarctic Glass Squid are most abundant in the Weddell Sea and Prydz Bay. Early life stages of paralarvae and juveniles suggest that the Prydz Bay region is a spawning ground. These young stages were found in epipelagic waters, while adults tend to live in mesopelagic and bathypelagic zones. When they die, they float to the surface, which makes them available to seabirds. Adults collected at the surface have often been attacked already and partially eaten, but can still provide valuable scientific information. For example, in 2003 the first mature female Antarctic Glass Squid ever described was found floating on the surface, and even though its head had already been consumed, its relatively undamaged mantle was packed with reproductive information.

Spermatangia, the packages of sperm left after spermatophores have discharged, were found on the female's mantle, showing that she had mated. Ripe eggs were in her oviducts, and several thousand unripe eggs were in her ovaries. It is not known how much of this "potential fecundity" would be realized, by spawning, as "actual fecundity." Other specimens have shown that unripe eggs are often resorbed. Scientists think the ability to resorb eggs may represent a flexible adjustment of female fecundity to body size. A high number of unripe eggs is available to her if she encounters excellent conditions and grows large enough to have the energetic resources to lay that many. However, if she never grows that big, she can resorb a number of eggs to reclaim the energy and lay a smaller number. (Humans, too, produce hundreds of thousands of unripe eggs, many of which disappear by the time we reach puberty.)

→   When first described in 1906, the Antarctic Glass Squid was given the genus name *Crystalloteuthis*, likely in references to its "glassy" crystalline appearance. Its tentacles are much longer than its itty-bitty arms, in sharp contrast to the next species we'll meet.

Antarctic New Squid

Short-tentacled South Pole specialist

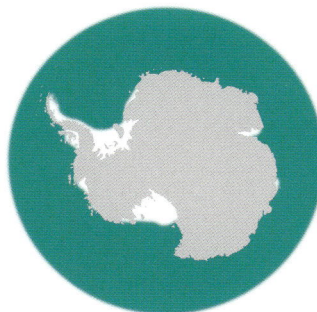

SCIENTIFIC NAME	*Alluroteuthis antarcticus*
FAMILY	Neoteuthidae
MANTLE LENGTH	10¾ in (27 cm)
TOTAL LENGTH	Estimated 3¼ feet (1 m)
NOTABLE ANATOMY	Short tentacles
MEMORABLE BEHAVIOR	Cannibalism

Found all around Antarctica, these cute little squid engage in both ontogenetic and diel migration. The tentacles of this species are not nearly as much longer than the arms as is typical for squid. They eat lanternfish, krill, and other squid (including their own species), and are preyed on in turn by elephant seals.

They appear to share a geographic area with Antarctic Glass Squid (*Galiteuthis glacialis*) by occupying different sub-regions, though it's not known if they compete for the same food or have also partitioned their diets. Both species participated (unintentionally) in a 2020 study about the quantity of mercury in squid in the Southern Ocean. Mercury contamination of seafood has long been a practical concern, as humans can be poisoned by eating shellfish with high concentrations of mercury. Of course, mercury can also damage the animals themselves, and becomes more problematic the higher it travels up the food chain. This is called bioaccumulation. Animals at the bottom of the food chain pick up a small amount of mercury from their environment, but they don't break it down or process it in any way; they simply store it in their tissues. As they become prey for other animals, those animals consume many small sources of mercury, similarly storing them, to become a greater source of mercury themselves. Squid, as predators,

could accumulate quite a lot of mercury this way, then pass it on up to everything that eats squid—dolphins and albatrosses and humans. Squid are an especially useful biological indicator of pollution such as mercury, because they grow fast and are short-lived. It's hard to get a real-time sense of pollution impacts on slow-growing, long-lived marine mammals, but squid respond to their environment right away.

Antarctic New Squid accumulated more mercury in their muscles as they grew but, curiously, Antarctic Glass Squid did the opposite. Overall, the study's news was encouraging—the animals had been collected over the decade from 2006 to 2016, and mercury concentrations were shown to decline over that time period. The authors concluded that mercury pollution in the Southern Ocean has been decreasing, with consequently reduced risks for predators. This is likely reflective of the effort that humans have made to reduce mercury input into the environment.

→   Unlike the Antarctic Glass Squid, the Antarctic New Squid has long arms and tentacles of a similar length. These adaptations could be related to differences in diet, which would allow these two species to coexist without competing, but such an idea remains speculative.

Antarctic Octopus

Short-armed southern muscle

SCIENTIFIC NAME	*Pareledone sp.*
FAMILY	Megaleledonidae
MANTLE LENGTH	4 in (10 cm)
TOTAL LENGTH	14 in (35 cm)
NOTABLE ANATOMY	Mantle papillae
MEMORABLE BEHAVIOR	Binding extra oxygen in its blood

There may be more than 20 distinct species of *Pareledone*, all found around Antarctica. Their musculature tends to be quite solid, and although they have no fins, they do have small internal shell vestiges called stylets. Their skin is covered with papillae that range from small and simple to large and branched. Species are distinguished by variation in the papillae, the pattern of white leucophore markings on their mantles, and the color of their eggs.

For most of the twentieth century there were very few specimens of *Pareledone* available, which made it difficult to figure out how many species there were. However, in the 1990s and early 2000s, trawls were conducted around Antarctica with commercial fishing gear that brought up thousands of *Pareledone* octopuses. This allowed researchers to begin identifying a range of features that merited sorting the genus into many more species than previously recognized. It was in this bonanza that the small delicate stylets were first discovered.

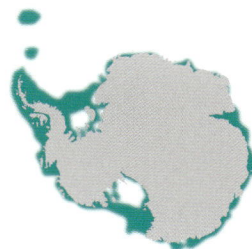

Male *Pareledone* octopuses produce spermatophores of astonishing length, up to twice as long as their mantles. Such a lengthy spermatophore must be produced in curved or coiled fashion. And as in the production of large eggs, only a few can be made at a time. Whereas the small spermatophores of other cephalopods may be stored in the hundreds, a representative *Pareledone* male had only five of these giant spermatophores stored in his body.

These octopuses are benthic, living on the continental shelf or slope in very cold water, down to 28 °F (−2 °C). One species, *P. charcoti*, was studied by scientists in 2015 to figure out how it copes with these low temperatures. Comparing its blood to that of octopuses from warmer water, they found that this Antarctic octopus had 40 percent more hemocyanin. This increase could make up for the oxygen-binding pigment's reluctance to let go of oxygen in cold water. Scientists also studied the hemocyanin itself, finding evidence that it can pass oxygen back and forth more efficiently at warmer temperatures—better than the hemocyanin of other species—which suggests that these octopuses may be able to tolerate a wide range of temperatures. Another species, Turquet's Octopus (*P. turqueti*), has been recorded from a wide depth range, 13,100 ft (4,000 m) all the way up to near the surface. They're clearly a flexible group.

→ Species in this genus display a lot of variety (different base colors, skin structures, presence or absence of ink sac) but also share certain features, such as a single row of suckers and males producing a spermatophore longer than their own body.

Arctic Octopus

North Pole worm-eater

SCIENTIFIC NAME	Cirroteuthis muelleri
FAMILY	Cirroteuthidae
MANTLE LENGTH	13 in (33 cm)
TOTAL LENGTH	5 ft (1.5 m)
NOTABLE ANATOMY	Large size
MEMORABLE BEHAVIOR	Eating worms

Living all around the Arctic, this *Cirroteuthis* is fairly substantial in size, perhaps an example of polar gigantism. Its eggs have a hard shell and are laid on the seafloor, where they develop for 20–32 months before hatching—a prime example of cold temperatures extending developmental duration. Females may also engage in brooding behavior, which is unexpected for a cirrate octopus. This species is most abundant around Alaska, Baffin Bay, and Davis Strait. It tends to keep some distance from the surface, living no shallower than 1,640 ft (500 m) and as deep as nearly 13,100 ft (4,000 m).

This was the first species in the genus *Cirroteuthis* to be described, when an individual was found off Greenland in 1838. Two more were found in 1846. In fact, these were the first cirrate octopuses of any kind to be described, and thus had the honor of introducing humans to the fact that there was a whole new kind of octopus, different from the ones we were used to, with cirri on their arms and fins on their mantles. New cirrate octopus species were all added to *Cirroteuthis* for a while, until the genus became encumbered enough that scientists began to divide it up.

As befits a classic cirrate octopus, these Arctic octopuses have very noticeable arm cirri, a very deep web that reaches nearly to their arm tips, large fins, and a sturdy internal shell. Their eyes are large, their body gelatinous. They eat crustaceans, which is typical octopus behavior, but they are also notable for their consumption of marine worms.

Cirroteuthis muelleri has been compared to the boreal Glowing Sucker Octopus (*Stauroteuthis syrtensis*) as an "ecological analogue." These two species occupy comparable niches in different locations. Both are fairly common, with scientists able to collect well over a hundred of each around Greenland, Iceland, and the Barents Sea. Other than *C. muelleri* occurring in more northern waters and *S. syrtensis* in more southern waters, they live at the same depths and similar densities of individuals. *Stauroteuthis syrtensis* lives at a toasty 38 °F (3.5 °C), while *C. muelleri* prefers a cooler 32 °F (0 °C). Estimates of abundance suggest an explanation for why cirrate octopuses were first discovered here—they appear to be far more densely populous than cirrate octopuses in warmer parts of the world.

→   This species has long cirri on its arms. Some of these are visible peeking out from below the octopus's arm web, which is also substantial.

ROSSIA PALPEBROSA

Warty Bobtail Squid

Arctic middle child

SCIENTIFIC NAME	*Rossia palpebrosa*
FAMILY	Sepiolidae
MANTLE LENGTH	1¾ in (4.5 cm)
TOTAL LENGTH	2¼ in (5.8 cm)
NOTABLE ANATOMY	Medium size
MEMORABLE BEHAVIOR	Medium depth and temperature preference

This species has what's known as a "nektobenthic" lifestyle, meaning that it hangs out near the bottom but often swims above the seafloor rather than sitting or crawling on it. Many studies of bobtail squids have identified individuals only to *Rossia* sp., which makes it difficult to know which traits apply to which species.

Rossia palpebrosa shares the Arctic with two other *Rossia* species: *R. megaptera* (Big-fin Bobtail Squid) and *R. moelleri*. With three such closely related species in the same habitat, how do they compete or share resources? It turns out that *R. palpebrosa* is the "middle child," having the widest temperature range and living at medium depth. *R. megaptera* is the smallest, lives at the deepest depth, and prefers the warmest water. *R. moelleri* is the largest, shallowest, and prefers the coldest water. These differences leave them still overlapping in habitat, and often multiple species are caught in the same location by the same method.

Scientists have been able to tease apart enough slight differences in the three species' diet and habitat to explain how they've all specialized to coexist. *Rossia moelleri*, for example, eats mostly fish, while the other two focus on crustaceans. *Rossia palpebrosa* remains in one location throughout its life, while the other two tend to migrate.

From 2007 to 2012, starting with the fourth International Polar Year, scientists made extensive collections of Arctic cephalopods. They were able to use shrimp bottom trawls to obtain 871 Warty Bobtail Squids from the Barents Sea and its vicinity. Using their collection data, knowledge of the area, and math, they calculated that the total number of Warty Bobtail Squid in the area ranges from 250 to 520 million individuals. The areas with the most abundance were in the depth range of 490–1,310 ft (150–400 m), and the temperature range of 34–32 °F (−1–0 °C). Warty Bobtail Squid are not evenly distributed, but concentrate in their preferred habitat, and although that sounds like many million individuals, it's not nearly as many as there are of other invertebrates and fish in the region. They're abundant for sepiolids, but not for Arctic animals.

The maximum abundance of this species during the study period was recorded in 2007 and 2012, which were also the warmest years on record. The researchers suggest that during these years, young Warty Bobtail Squid got more food thanks to the higher temperatures (which usually correlate with a greater mass of animals on the seafloor) and more of them were able to survive to adulthood as a result.

→ These cute little bobtails are thought to eat primarily crustaceans throughout their range, followed in dietary importance by fish and worms.

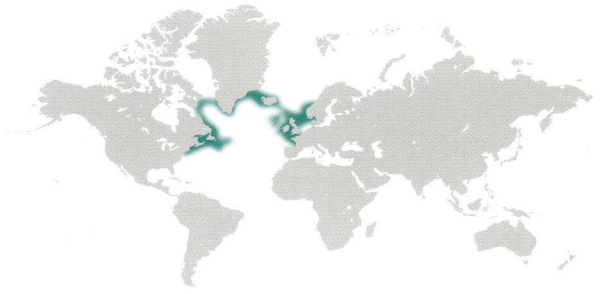

BATHYPOLYPUS ARCTICUS

North Atlantic Spoonarm Octopus

Proud papa

SCIENTIFIC NAME	*Bathypolypus arcticus*
FAMILY	Bathypolypodidae
MANTLE LENGTH	2¾ in (7 cm)
TOTAL LENGTH	9 in (23 cm)
NOTABLE ANATOMY	Warty skin
MEMORABLE BEHAVIOR	Curling arm tips

Despite their scientific name, these octopuses are only edging into true Arctic habitat. They are found off northern Canada, Greenland, and northern Europe. They usually live greater than 1,310 ft (400 m) deep, and seem inclined to shallower water the further north they are. Large skin papillae sometimes called "warts" dot their skin, and the reproductive arm of mature males ends with a large flattened tip like the bowl of a spoon, giving them their name. They have no ink sac.

North Atlantic Spoonarm Octopuses have been successfully kept in the laboratory, to the point of laying eggs that hatched. The researchers who studied them remarked on the sturdiness of the species, which allowed individuals to survive a rather traumatic unintentional collection. Eighteen octopuses were found as accidental bycatch of a scallop-collecting expedition in Canada's Bay of Fundy.

They mated readily in captivity, with males wrapping females in their arm web and sticking their hectocotylus inside the female's mantle. Male North Atlantic Spoonarm Octopuses have an enormous ligula, part of their hectocotylus. Scientists guess that the hectocotylus is inserted all the way into the oviduct, where it must make a significant effort to deliver the large spermatophore. The size of the spermatophore may be an adaptation to rare encounters in the wild, favoring individuals who make the most of each mating opportunity. Bigger spermatophores may be better able to compete with or even flush out those of other males.

Brooding in captivity lasted more than a year, and the octopus's life span is thought to be about three years, perhaps reaching up to a venerable six years—a range that is similar to the estimated life spans of Giant Pacific Octopus (*Enteroctopus dofleini*), but far longer than the typical life span of an octopus as small as the Spoonarm. Rather than adhering to a strict fast, the females would occasionally accept food offered by researchers. However, they remained with their eggs, aerating them with water from the funnel, and attacked the scientists' probe with arms or jets of water when it approached. Eating didn't stop them from dying, some before and some after their eggs hatched. Those that died before hatching moved away from their eggs to die.

Video of hatching North Atlantic Spoonarm Octopus babies is impossibly endearing. Many hatching octopuses that have been recorded on video are paralarvae, squirting out of their eggs and then drifting away on the currents. But the Spoonarm Octopus wriggles free of its egg case, its arms reaching immediately for the seafloor, tips curled up in characteristic posture, and then toddles off.

→ These octopuses, like many deep-sea species, lack ink sacs. Their bodies are similar to Antarctic benthic octopuses, with relatively short arms, partial webs, and sturdy mantles.

GLOSSARY

abyssal Oceanic habitat between 13,100 and 19,700 ft (4,000 and 6,000 m), including open water (abyssopelagic) and seafloor (benthic), or organisms occupying this habitat.

anthropogenic Produced by humans, often used to describe environmental impacts such as pollution or climate change.

aposematism A visual display to warn predators that the animal bearing it is toxic or otherwise dangerous.

arm A cephalopod appendage lined with suckers from base to tip; octopuses, squid, and cuttlefish all have eight arms.

autotroph An organism that produces its own food from an inorganic energy source, such as sunlight or chemicals.

axon The elongate part of a nerve cell that conducts signals to other cells; squid have giant axons that facilitate their escape jets.

Batesian mimicry A visual display in a non-threatening species that mimics the aposematism of a dangerous species, invoking the same predator aversion.

bathyal Oceanic habitat between 3,300 and 13,100 ft (1,000 and 4,000 m), including open water (bathypelagic) and seafloor (benthic), or organisms occupying this habitat.

benthic Marine habitat on the seafloor at any depth, or organisms occupying this habitat.

biomass The mass of living organisms, often of a particular species or group of interest, found per unit area.

biomimetics A field of engineering that takes inspiration from living organisms to create materials, robots, buildings, etc.

buoyancy The force of surrounding fluid that lifts an object in opposition to gravity, also used to describe a property of the object itself, where an object with positive buoyancy rises, with negative buoyancy sinks, and with neutral buoyancy remains in position.

chemosynthesis The production of organic sugars from inorganic chemical energy, such as performed by bacteria at hot vents and cold seeps.

chitin A biological material similar to the keratin in human fingernails, used by cephalopods to build beaks and internal shells.

chromatophore An organ in the skin of cephalopods comprising nerves, muscles, and pigment, which can function as a "pixel," turning on or off to contribute to skin color patterns.

cirrate The group of octopuses with paired fins and thin skin protrusions (cirri) lining their arms along with suckers; the group of octopuses without these features are **incirrate**.

coleoid The group of cephalopods that evolved internal shells; modern representatives include squid, cuttlefish, and octopuses.

commensalism A biological interaction of two or more species living together, without the hallmarks of mutualism or parasitism.

convergent evolution The process of organisms from different groups evolving similar traits in response to similar environments.

consort male A male in some squid and cuttlefish species which competes with other males for access to mates and guards mates from mating attempts by other males.

counterillumination The production of light via bioluminescence on the underside of an organism, which helps it blend with the well-lit surface to an observer looking up from below.

cuttlebone The internal shell of a cuttlefish, made of calcium carbonate and containing gas-filled chambers used to regulate buoyancy.

deep scattering layer The marine organisms, largely fish and crustaceans but including squid, that aggregate densely enough to reflect sonar and migrate to the surface at night and to depth during the day.

deimatic display A visual display that can startle or scare off potential predators.

diel migration The behavior of aquatic animals moving to shallower depths at night to feed and deeper depths during the day to hide.

disruptive coloration A pattern that breaks up the natural shape of an organism, making it difficult to recognize either in whole or in part, such as a disruptive eye mask obscuring the eyes.

doratopsis The distinctive paralarval stage of chiroteuthid squid, possessing a long decorated tail that resembles a siphonophore.

ecosystem A particular habitat and all the organisms that inhabit it and interact with one another.

ecosystem engineer A living organism, such as reef-building corals, that creates the physical structure of an ecosystem.

endosymbiosis An interaction in which one organism lives completely inside another, often practiced by microbes inside animals and plants.

epipelagic Oceanic habitat between the surface and 660 ft (200 m), or organisms occupying this habitat.

ethogram A catalog of an animal's behaviors created by scientists.

gape The size, typically delimited by an animal's jaw, that constrains the size of prey it can take, which in cephalopods is delimited by the arm crown rather than beak.

genome The complete set of genes or DNA in an organism.

gladius A stiff but flexible internal rod that is the shell remnant of squid, vampire squid, and various coleoid ancestors.

hadal Oceanic habitat below 19,700 ft (6,000 m), including open water (hadalpelagic) and seafloor (benthic), or organisms occupying this habitat.

hectocotylus The reproductive arm of many male coleoids, modified for sperm delivery and detachable, in the case of argonauts.

hemocyanin The cephalopod respiratory pigment, copper-based by contrast with the iron-based hemoglobin in many other animals, making their blood blue rather than red.

holobiont The entity that consists of all symbionts sharing a given space; often refers to a host and its microbial communities.

intertidal Habitat between low tide mark and high tide mark, alternately submerged in seawater and exposed to air, or organisms occupying this habitat.

iridophore A reflective organ in the skin of cephalopods that can function as a "pixel," turning on or off to contribute to skin color patterns.

iteroparity Reproductive feature of a species in which adults can spawn multiple times, continuing to feed between spawning events.

keystone species A species which occupies such an important position in a given ecosystem that, in its absence, the ecosystem is significantly altered.

leucophore An organ in the skin of cephalopods that reflects ambient light, providing a "background" for skin color patterns.

mantle The main body or "torso" of a cephalopod that contains its organs, squeezes out water to produce jet propulsion, and secretes the shell if present.

marine snow Fragmentary detritus that sinks in the ocean, including fecal pellets, discarded molts, and dead plankton.

mesopelagic Oceanic habitat between 660 and 3,300 ft (200 and 1,000 m), or organisms occupying this habitat.

microbe A single-celled organism, including bacteria and fungi, many of which live as endosymbionts of multicellular organisms.

morphology The physical forms of living organisms and the study thereof.

mutualism A biological interaction of two or more species living together in which all participants derive benefits.

neoteny An evolutionary process whereby features of larvae or juveniles are retained into adulthood.

neritic Shallow marine habitat near coasts, or organisms occupying this habitat.

niche An organism's habitat, food source, and interactions with other organisms.

ontogenetic The development of an animal from embryo through adult, during which many marine animals migrate from one habitat to another.

osmosis The process of fluid crossing a membrane from the area of greater solute concentration to the area of lesser solute concentration, which can create osmotic stress in freshwater organisms.

oxygen minimum zone (OMZ) A region in the ocean where oxygen is depleted due to the decomposition of organic material, often from 660 ft to 3,300 ft (200 to 1,000 m) depth.

papillae Structures in cephalopod skin that can be raised or lowered by muscles to contribute to camouflage or other displays.

paralarvae The early life stages of many cephalopods that are morphologically distinct from adults but do not metamorphose like true larvae.

parasitism A biological interaction of two or more species living together in which one participant benefits at the expense of another.

pelagic Marine habitat in the open ocean, not near the coast, or organisms occupying this habitat.

pen See *gladius*.

photophore An organ that can produce light, often thanks to symbiosis with bioluminescent bacteria, used by cephalopods for camouflage and communication.

photoreceptor Cells that convert light into a nerve signal, present both in the eyes and in the skin of cephalopods.

plankton Aquatic organisms that drift with the currents, typically but not always very small in size; the foundation of open ocean food webs.

polarization The phenomenon of light waves vibrating in a single plane, a feature than can be detected by cephalopod and stomatopod (mantis shrimp) eyes but not human eyes.

pycnocline A transitional layer in the ocean between less-dense surface water and denser deep water.

radula A mollusk mouthpart, similar to a toothed tongue, which can be used to drill holes, to scrape flesh, and as a conveyor belt to move food down the esophagus.

rhynchoteuthion The distinctive paralarval stage of flying squid, with the two tentacles fused into a proboscis.

ROV A remotely operated underwater vehicle used to explore the ocean, unoccupied by humans and tethered to a ship.

salinity A measure of the amount of salt dissolved in water, typically around 35 parts per thousand in the ocean.

semelparity Reproductive feature of a species in which adults can spawn only once, then die.

senescence The process of aging at the end of an organism's life.

sexual dimorphism Difference in physical form (color, shape, size, or other) between the sexes of a given species.

siphonophore A colonial marine organism related to jellyfish, with members of the colony specialized to feed, float, sting, or reproduce.

sneaker male A male in some squid and cuttlefish species that, rather than competing with other males for access to mates, attempts to mate with females that are already being guarded.

spermatophore A complex reproductive package produced by male cephalopods, consisting of a sperm mass and ejaculatory apparatus.

squad A group of squid swimming together.

statocyst A cephalopod's balance organ, consisting of a hard structure called a statolith inside a chamber lined with sensitive cells.

substrate A surface or material on which an organism lives or moves, such as sand or rock.

symbiosis A biological interaction of two or more species living together, in which participants may or may not benefit.

tentacle A cephalopod appendage that is usually retractile and has suckers on the tip (in coleoids) or is sticky and can be moved in and out of a sheath (in nautiluses).

thermocline A transitional layer in the ocean between warm surface water and cool deep water.

thermohaline The circulation pattern of the global ocean, driven by temperature and salinity.

upwelling The process of deeper, colder, and more nutrient-rich water being drawn toward the surface of the ocean to replace shallower, warmer water, when winds push the latter offshore.

→ In aquariums around the world, we ask octopuses to take our pictures, set them up on dates for Valentine's Day, and have them predict soccer match outcomes. Let us also love them for themselves.

RESOURCES

BOOKS

Boyle, P. and P. Rodhouse. *Cephalopods: Ecology and Fisheries* (Blackwell Science, 2005)

Godfrey-Smith, P. *Other Minds: The Octopus, the Sea, and the Deep Origins of Consciousness* (Farrar, Straus and Giroux, 2016)

Hanlon, R.T. and J.B. Messenger. *Cephalopod Behaviour*. 2nd edition (Cambridge University Press, 2018)

Hanlon, R., M. Vecchione, and L. Allcock. *Octopus, Squid, and Cuttlefish: A Visual, Scientific Guide to the Oceans' Most Advanced Invertebrates* (University of Chicago Press, 2018)

Jereb, P. and C.F.E. Roper, eds. *Cephalopods of the World: An Annotated and Illustrated Catalogue of Cephalopod Species Known to Date, Volume 1: Chambered Nautiluses and Sepioids* (Food and Agriculture Organization of the United Nations, 2005)

Jereb, P. and C.F.E. Roper, eds. *Cephalopods of the World: An Annotated and Illustrated Catalogue of Cephalopod Species Known to Date, Volume 2: Myopsid and Oegopsid Squids* (Food and Agriculture Organization of the United Nations, 2010)

Jereb, P., C.F.E. Roper, M.D. Norman, and J.K. Finn, eds. *Cephalopods of the World: An Annotated and Illustrated Catalogue of Cephalopod Species Known to Date, Volume 3: Octopods and Vampire Squids* (Food and Agriculture Organization of the United Nations, 2016)

Mather, J.A., R.C. Anderson, and J.B. Wood. *Octopus: The Ocean's Intelligent Invertebrate* (Timber Press, 2010)

Montgomery, S. *The Soul of an Octopus: A Surprising Exploration into the Wonder of Consciousness* (Atria Books, 2016)

Norman, M.D. *Cephalopods: A World Guide* (ConchBooks, 2000)

Okutani, T. *Cuttlefishes and Squids of the World* (new edition). (Publication of the 50th Anniversary of the Foundation of National Cooperative Association of Squid Processors, 2015)

Saunders, W.B. and N.H. Landman. *Nautilus: The Biology and Paleobiology of a Living Fossil* (Springer, 2009)

Staaf, D. *Monarchs of the Sea: The Extraordinary 500-Million-Year History of Cephalopods* (The Experiment, 2020)

Williams, W. *Kraken: The Curious, Exciting, and Slightly Disturbing Science of Squid* (Abrams, 2011)

SCIENTIFIC JOURNAL ARTICLES

Allcock, A.L., J.M. Strugnell, H. Ruggiero, and M.A. Collins. "Redescription of the deep-sea octopod *Benthoctopus normani* (Massy 1907) and a description of a new species from the Northeast Atlantic." *Marine Biology Research* 2 (6): 372–387 (2007). https://doi.org/10.1080/17451000600973315

Barratt, I.M., M.P. Johnson, and A.L. Allcock. "Fecundity and reproductive strategies in deep-sea incirrate octopuses (Cephalopoda: Octopoda)." *Marine Biology* 150: 387–398 (2007). https://doi.org/10.1007/s00227-006-0365-6

Boal, J.G. and S.A. Gonzalez. "Social behaviour of individual oval squids (Cephalopoda, Teuthoidea, Loliginidae, *Sepioteuthis lessoniana*) within a captive school." *Ethology* 104 (2): 161–178 (2010). https://doi.org/10.1111/j.1439-0310.1998.tb00059.x

Bowers, J., T. Nimi, J. Wilson, S. Wagner, D. Amarie, and V. Sittaramane. "Evidence of learning and memory in the juvenile dwarf cuttlefish *Sepia bandensis*." *Learning & Behavior* 48 (4): 420–431 (2020). https://doi.org/10.3758/s13420-020-00427-4

Burford, B.P., B.H. Robison, and R.E. Sherlock. "Behaviour and mimicry in the juvenile and subadult life stages of the mesopelagic squid *Chiroteuthis calyx*." *Journal of the Marine Biological Association of the United Kingdom* 95 (6): 1221–1235 (2015). https://doi.org/10.1017/S0025315414001763

Golikov, A.V., F.R. Ceia, R.M. Sabirov, G.A. Batalin, M.E. Blicher, B.I. Gareev, G. Gudmundsson et al. "Diet and life history reduce interspecific and intraspecific competition among three sympatric Arctic cephalopods." *Scientific Reports* 10 (1): 21506 (2020). https://doi.org/10.1038/s41598-020-78645-z

Grearson, A.G., A. Dugan, T. Sakmar, D.M. Sivitilli, D.H. Gire, R.L. Caldwell, C.M. Niell, G. Dölen, Z. Yan Wang, and B. Grasse. "The lesser Pacific striped octopus, *Octopus chierchiae*: an emerging laboratory model." *Frontiers in Marine Science* 8 (2021). www.frontiersin.org/article/10.3389/fmars.2021.753483

Hall, K. and R. Hanlon. "Principal features of the mating system of a large spawning aggregation of the giant Australian cuttlefish *Sepia apama* (Mollusca: Cephalopoda)." *Marine Biology* 140 (March): 533–545 (2002). https://doi.org/10.1007/s00227-001-0718-0

Hanlon, R.T., A.C. Watson, and A. Barbosa. "A 'mimic octopus' in the Atlantic: flatfish mimicry and camouflage by *Macrotritopus defilippi*." *The Biological Bulletin* 218 (1): 15–24 (2010). https://doi.org/10.1086/BBLv218n1p15

Hoving, H.J.T. and B.H. Robison. "Vampire squid: detritivores in the oxygen minimum zone." *Proceedings of the Royal Society B: Biological Sciences* 279 (1747): 4559–4567 (2012). https://doi.org/10.1098/rspb.2012.1357

Hoving, H.J.T. and B.H. Robison. "The pace of life in deep-dwelling squids." *Deep Sea Research Part I: Oceanographic Research Papers* 126 (August): 40–49 (2017). https://doi.org/10.1016/j.dsr.2017.05.005

Huffard, C.L., N. Saarman, H. Hamilton, and W. Brian Simison. "The evolution of conspicuous facultative mimicry in octopuses: an example of secondary adaptation?" *Biological Journal of the Linnean Society* 101 (1): 68–77 (2010). https://doi.org/10.1111/j.1095-8312.2010.01484.x

Jamieson, A.J. and M. Vecchione. "First in situ observation of Cephalopoda at hadal depths (Octopoda: Opisthoteuthidae: *Grimpoteuthis* sp.)." *Marine Biology* 167 (6): 82 (2020). https://doi.org/10.1007/s00227-020-03701-1

Kurita, Y. "Biological report of a giant deep-sea squid *Onykia robusta* collected from the Sanriku Coast, Japan: implications for low genetic diversity." *Marine Biodiversity* 48 (1): 685–688 (2018). https://doi.org/10.1007/s12526-017-0770-8

Montague, T.G., I.J. Rieth, S. Gjerswold-Selleck, D. Garcia-Rosales, S. Aneja, D. Elkis, N. Zhu et al. "A brain atlas of the camouflaging dwarf cuttlefish, *Sepia bandensis.*" *bioRxiv* (2022). https://doi.org/10.1101/2022.01.23.477393

Montana, J., J.K. Finn, and M.D. Norman. "Liquid sand burrowing and mucus utilisation as novel adaptations to a structurally-simple environment in *Octopus kaurna* Stranks, 1990." *Behaviour* 152 (14): 1871–1881 (2015). https://doi.org/10.1163/1568539X-00003313

Oellermann, M., B. Lieb, H.O. Pörtner, J.M. Semmens, and F.C. Mark. "Blue blood on ice: modulated blood oxygen transport facilitates cold compensation and eurythermy in an Antarctic octopod." *Frontiers in Zoology* 12 (1): 6 (2015). https://doi.org/10.1186/s12983-015-0097-x

Seco, J., J.C. Xavier, A.S. Brierley, P. Bustamante, J.P. Coelho, S. Gregory, S. Fielding et al. "Mercury levels in Southern Ocean squid: variability over the last decade." *Chemosphere* 239 (January): 124785 (2020). https://doi.org/10.1016/j.chemosphere.2019.124785

Seibel, B.A., N.S. Häfker, K. Trübenbach, J. Zhang, S.N. Tessier, H.O. Pörtner, R. Rosa, and K.B. Storey. "Metabolic suppression during protracted exposure to hypoxia in the jumbo squid, *Dosidicus gigas*, living in an oxygen minimum zone." *Journal of Experimental Biology* 217 (14): 2555–2568 (2014). https://doi.org/10.1242/jeb.100487

Sinn, D.L. and N.A. Moltschaniwskyj. "Personality traits in dumpling squid (*Euprymna tasmanica*): context-specific traits and their correlation with biological characteristics." *Journal of Comparative Psychology* 119 (1): 99–110 (2005). https://doi.org/10.1037/0735-7036.119.1.99

Strugnell, J.M., A.D. Rogers, P.A. Prodöhl, M.A. Collins, and A.L. Allcock. "The thermohaline expressway: the Southern Ocean as a centre of origin for deep-sea octopuses." *Cladistics* 24 (6): 853–860 (2008). https://doi.org/10.1111/j.1096-0031.2008.00234.x

Thomas, K.N., B.H. Robison, and S. Johnsen. "Two eyes for two purposes: *in situ* evidence for asymmetric vision in the cockeyed squids *Histioteuthis heteropsis* and *Stigmatoteuthis dofleini*." *Philosophical Transactions of the Royal Society B: Biological Sciences* 372 (1717): 20160069 (2017). https://doi.org/10.1098/rstb.2016.0069

Tramacere, F., A. Kovalev, T. Kleinteich, S.N. Gorb, and B. Mazzolai. "Structure and mechanical properties of *Octopus vulgaris* suckers." *Journal of The Royal Society Interface* 11 (91) (2014). https://doi.org/10.1098/rsif.2013.0816

Vecchione, M. "ROV observations on reproduction by deep-sea cephalopods in the central Pacific Ocean." *Frontiers in Marine Science* 6 (2019). www.frontiersin.org/article/10.3389/fmars.2019.00403

Von Byern, J. and W. Klepal. "Adhesive mechanisms in cephalopods: a review." *Biofouling* 22 (5): 329–338 (2006). https://doi.org/10.1080/08927010600967840

ORGANIZATIONS DEDICATED TO THE STUDY AND APPRECIATION OF CEPHALOPODS

Association for Cephalopod Research (CephRes)
www.cephalopodresearch.org

Cephalopod International Advisory Council (CIAC)
https://cephalopod.wordpress.com

OctoNation (The Largest Octopus Fan Club)
https://octonation.com

The Octopus News Magazine Online (TONMO)
https://tonmo.com

INDEX

PICTURE CREDITS

The publisher would like to thank the following for permission to reproduce copyright material:

AdobeStock/ pipehorse 53; /Daizen Takahashi 156; /EyeEm /anita van der mespel 173; /zampa 71.

AGE Fotostock/Marevision/Ignacio González Vázquez 265.
Alamy Stock Photo/Adisha Pramod 235; /Andrew Trevor-Jones 75; /Anthony Pierce 27;/cbimages 109; /David Fleetham 215t; /Helmut Corneli 12-13; /imagegallery2 158; /Images & Stories 73; /Jacob Maentz 55; /Lorne Chapman/Stockimo 155; /Mark Conlin 22l; /Nature Picture Library/Alex Mustard 63; /Nature Picture Library/David Shale 239; /Nature Picture Library/Magnus Lundgren 153; /Nobuo Matsumura 148r; /Richardom 69; /Rosanne Tackaberry 51r; /Sabena Jane Blackbird 21t; /Sipa US 224.
© **Atlantic Productions**/Discovery 2020 from Five Deeps for Discovery, photograph by Tamara Stubbs 216.
BluePlanetArchive.com/Andrew J. Martinez 154; /Jeff Milisen 175; / Mark Conlin 97l; /Andy Murch 229 & 243; /Christopher Swann 182-183; /Clay Bryce 91l; /Dave Forcucci 184; /David B. Fleetham 125; /David Wrobel 185, 211, 231b; /Doug Perrine 171; /Franco Banfi 188; /Gerald Nowak 139; /John C. Lewis 56; /Kat Bolstad 197 inset; /Marilyn & Maris Kazmers 91r; /Mark Strickland 39; /Masa Ushioda 148l; /Michael Valos 217; /Solvin Zankl 181; /Steven Kovacs 122l, 149, 169.
Dr. Roy Caldwell/Department of Integrative Biology, University of California, Berkeley 101.
Cassell's natural history Vol 5, Peter Martin Duncan, 1821-1891 160.
Andrew D. Corso/Fisheries Science Department, Virginia Institute of Marine Science 269.
DESCNA Deep Sea Creatures of the North Atlantic, Brønshøj, Denmark 275.
Die Cephalopoden by Carl Chun, vol II. Teil: Myopsida, Octopoda, from "Wissenschaftliche Ergebnisse der Deutsche Tiefsee-Expedition auf dem Dampfer "Valdivia" 1898-1899. Bd. XVIII. 2. Teil." Verlag von Gustav Fisher in Jena, 1915 263l&r.
Die Cephalopoden der Plankton-Expedition by Georg Pfeffer, Ergebn. Planktonexped. Pt. 2 (vol 3 plates). Kiel & Leipzig: Lipsius & Tischer. 1912, Plates 45 & 46. 192 & 193.
Dreamstime.com/Mario Pesce 97r; /Jaahnlieb 83b; /John Anderson 135; /Shermix 52r.
Flickr/wenfisher-(CC BY 2.0) 143.
Jan van Franeker/Wageningen Marine Research 267.
Gettyimages/Paul Starosta 194b.
Dr Sam Glenn-Smith (@sammyglenn_dives) 67.
imagequestmarine.com/Peter Batson 237.
istockphoto/Michael Zeigler 94.
Tim Karnasuta (@timkarnasuta) 103.
Uwe Kils (CC-BY-SA-3) 257c.
Douglas Klug Santa Barbara, CA klugd@cox.net 83t.
Dr. Lesanna Lahner, Wildlife Veterinarian 198.
Ministry of Fisheries (CC BY) 197.
Monterey Bay Aquarium Research Institute/© 2004 MBARI 29c & 186; ©2020 MBARI 233b.
NASA/Goddard Space Flight Center Scientific Visualization Studio 252-253; 256; /Michael Studinger 255t.
Nature in Stock/Minden Pictures/Fred Bavendam 165; /Norbert Wu 260l & r & 262; /Oceanwide/Gary Bell 40.
naturepl/Alex Mustard 99 & 137; /David Shale 277; /Jeff Rotman 41; /Nature Production/Noriaki Yamamoto 203; /Solvin Zankl 11, 201, 228 & 231t.
NOAA/CCMA Biogeography Team 81c; / Office of Ocean Exploration and Research 225t & 227t' / Okeanos Explorer Program 215b, 221t; 222c 226t, 227b & 233t; /Vents Program/Image courtesy of Submarine Ring of Fire 2004 Exploration 218; /NOS-ORR 146c; /CINMS /Sanctuaries-Channel Islands NMS 82; /Courtesy of Dr. John R. Dolan,

Laboratoire d'Oceanographique de Villefranche; Observatoire Oceanologique de Villefrance-sur-Mer 147; /Monterey Bay Aquarium Research Institute 226b; /NPS Submerged Resources Center, Photo by Brett Seymour 79; /OAR/OER/Kevin Raskoff, Monterey Peninsula College/ Hidden Ocean Expedition 2005 194t; /Olympic Coast NMS; NOAA/OAR/Office of Ocean Exploration. Photo by Ed Bowlby 29t.
NTNU University Museum (CC BY-SA 4.0) 205.
OceanwideImages.com/Gary Bell 113.
Karen J. Osborn/Smithsonian 209.
Pixabay/Alice Key Studio 116; /d_alexander33 120; /LivingFlyLegacy 87.
Kevin Raskoff/Monterey Peninsula College 273.
Report on the Scientific Results of the Voyage of H.M.S. Challenger, during the Years 1873-76 under the command of Capt George S. Nares R.N. F.R.S. and Capt Frank Tourle Thomson, R.N. Vol 1. Physics and Chemistry, published London, Longmans & Co., John Murray; Macmillan & Co et al 214.
Résultats des campagnes scientifiques accomplies sur son yacht par Albert Ier, prince souverain de Monaco by Louis Joubin (1861–1935), Vol 17 Céphalopodes Plate 14. 1900 195.
Marcia Riederer (@marciariederer) 107.
Photo courtesy of Schmidt Ocean Institute 163 & 190.
Science Photo Library/British Antarctic Survey 271; /Dante Fenolio 207, 245 & 247; /Eye Of Science 22r.
SeacologyNZ.com/Crispin Middleton 167.
Shutterstock/Agarianna76 65; /Alex GK Lee 50-51c; /Alex Permiakov 122r; /Anton Ivanov 250; /aurelie le moigne 96; /azure 46; /Bilanol 264; /BlueBarronPhoto 252l; /bluehand 4br; /Boban Vaiagich 281; /Brent Barnes 111; /Charlotte Bleijenberg 28l; /Choksawatdikorn 146b; /Cingular 118; /Citrus deliciosa 86; /Colin Harnish 259t; /Dario Sabljak 31; /Diego Barucco 21b; /Divelvanov 2, 92-93 & 129r; /Dogora Sun 44; /Dustie 58; /elena_photo_soul 47; /Ethan Daniels 119 & 121t; /George P Gross 90; /Gerald Robert Fischer 4cr; /Globe Guide Media Inc 49; /Henner Damke 3 & 19; /Howard Noel 85; /Joe Belanger 123; /John Back 18; /Josef Hanus 48; /lavizzara 84; / luchschenF 259c; /magnusdeepbelow 35; /Malshak 89b; / marcobrivio.photo 78; /Marla_Sela 50l; /Mike Workman 34r, 128c & 129l; /Morphart Creation 10; /Noa Siti Eliyahu 16; /Oksana Golubeva 131; /Paul Vinten 33; /Pecold 45; /Peter Douglas Clark 26; /Peter Leahy 127; /Philip Garner 105; /Pierre-Yves Babelon 81t; /randi_ang 37r; /Reimar 121b; /Rich Carey 4tl, 88, 117b & 126; /Richard Whitcombe 92 & 157t; /Sakis Lazarides 9 & 128b; /SaltedLife 161; /Sam Robertshaw 5t&b;/Sarah2 28r; /Sascha Janson 141; /SergeUWPhoto 117t; /Sergey Muhlynin 89r; /Shiva N hegde 52l; /Shpatak 37l & 150; /Sineenuch J 57; /Solodov Aleksei 4tr & 255b; /Squidshooting 38; /Sven Philipp 60; /Tarpan 187; /Thierry Eidenweil 34l; /V.Gordeev 225b; /Visions-AD 162; /Vladimir Turkenich 24; /vovidzha 159; /wildestanimal 95 & 157b; /Wirestock Creators 257t; /xolmgard 251; /Yuri-D3 219; /Yusran Abdul Rahman 130.
Peter B Southwood 32.
Becky Tooby, Pembroke, Wales 133.
Steven Haddock/MBARI 177 & 241.
The North Pole. Its discovery in 1900 under the auspices of the Peary Arctic Club by Robert E. Peary. 2nd edition, Frederick A. Stokes Company, New York, 1910 258.
Courtesy of University of Washington, NSF/Ocean Observatories Initiative/Canadian Scientific Submersible Facility, CC-BY-SA-2 222l.
Woods Hole Oceanographic Institution, D. Bewley 191.

All reasonable efforts have been made to trace copyright holders and to obtain their permission for the use of copyright material. The publisher apologizes for any errors or omissions in the list above and will gratefully incorporate any corrections in future reprints if notified.